G. HASENJAEGER

INTRODUCTION TO THE BASIC CONCEPTS AND PROBLEMS OF MODERN LOGIC

D. REIDEL PUBLISHING COMPANY / DORDRECHT-HOLLAND

EINFÜHRUNG IN DIE GRUNDBEGRIFFE UND PROBLEME DER MODERNEN LOGIK

First published in 1962 by Verlag Karl Alber, Freiburg/München
Translated from the German by E. C. M. Mays

Library of Congress Catalog Card Number 70–135108

ISBN 90 277 0206 3

Printed in The Netherlands by D. Reidel, Dordrecht

2pcg
20c/0 LCO 08/05

INTRODUCTION TO THE BASIC CONCEPTS
AND PROBLEMS OF MODERN LOGIC

PREFACE

The field of modern logic is too extensive to be worked through by open-cast mining. To open it up, we need to sink shafts and construct adits. This is the method of most text books: a systematic exposition of a number of main topics, supplemented by exercises to teach skill in the appurtenant techniques, lays a secure foundation for subsequent discussion of selected questions.

Compared with this, the present treatment is more like a network of exploratory drillings to show that it would be worthwhile to start mining operations, or to work the existing shafts and adits, as the case may be. Within this metaphor we may also describe the inherent weakness of this conception: once a cavity is pierced, the duct's capacity will in general not be sufficient to carry away the discovered riches. But whether we are concerned with a new or an already worked mine – at any rate, the experience should stimulate us into either reviving an existing system of shafts or even, in particularly fortunate cases, designing a new approach.

Discarding our metaphor: brief accounts, of some of the various aspects of logic, will have served their purpose if they give the incentive to a more thorough study of some of the questions thrown up by these aspects. Sooner or later this will necessitate falling back on systematic expositions. However, in my view, there are worse ways of preparing for such reading than to gain a first-hand experience beforehand, through one's own intellectual efforts, of the questions that such reading will raise. The hints contained in Sections III 3 and IV 2, in particular, are intended to contribute towards this.

Because of its arrangement according to aspects, this introduction is no example of a deductive and logically self-contained exposition of a branch of knowledge. Even less is it a text book of some elementary part of logic, whose acquisition could be regarded as a prerequisite condition of all scientific endeavour. In my opinion, this kind of logic is acquired not *before* but *with* the knowledge in question, and we start on this process already when we learn to speak. The way in which we reach awareness of logical laws, will concern us later (p. 9 ff.).

In keeping with the aim of this book – that of motivating concepts of modern logic and problems linked with them – I have chosen the following arrangement: the first half deals with so-called first-order predicate logic (as the kernel of modern logic), this being (I) analysed as a further development of traditional logic, (II) explained as theory of (really: as a surveyable discrete section from) the existent, whose language reflects so much of the latter's structure that it (III) can substitute for the existent as object of investigation (or has our 'discrete ontology' suggested itself to us only because of the necessarily discrete structure of language?). Finally in (IV) we examine the correspondence between linguistic-deductive and ontic-relational structure.

The second half presents (V) practical and fundamental extensions of the language of logic (and hence of the world picture afforded by it) and draws attention to the openness of all (VI) expressive and (VII) inferential means in this extended domain. This *trivium* of the non-trivial is supplemented by a look at (VIII) the logic of probability, which cannot be included under the extensions discussed earlier (V).

The translator, Mrs. E. C. M. Mays, also deserves to be thanked for her help in adapting several linguistic examples as well as in proof-reading. I have used the occasion of this translation to improve some passages, mostly to avoid ambiguities of the German edition. As for the bibliography: whenever an item was quoted in the German edition for further reading rather than for particular reference, I tried to replace it by a comparable, possibly newer, source written in English.

Bonn, August 1971 G. HASENJAEGER

TABLE OF CONTENTS

CHAPTER I

TRADITIONAL AND MODERN LOGIC

It is quite possible to have learned to make 'logically correct' inferences without having explicit knowledge of any of the rules of logic. The special discipline of logic has come into being only as the result of a conscious search for rules of inference and their explicit formulation. These rules may be expressed in very different ways, e.g.:

(1) by collecting typical examples of any one mode of inference;

(2) by the description, in words, of the structure or form of a mode of inference, possibly with the use of 'variables';

(3) by the representation of the underlying rule mainly or exclusively in terms of a suitable symbolic notation or conceptual language.

The first method has the advantage of being intuitive and easily remembered. However, it is not always clear just what degree of generality of the underlying rule the examples are intended to convey, or is attached to them by the speaker or listener. If all the examples relate to one topic – which would in itself be of no relevance in this connexion – the reader will often have difficulty in applying the rules correctly to other subject matters. This method is comparable to that of the sage who answers questions regarding the nature of the good by telling stories. In many cases this method leads to a much clearer representation of the idea of the good than would a definition – if, indeed, there is such a definition. However, in the case of logic we need to do more than tell stories.

The second method is represented by a number of systems of logical laws dating from the time of classical Greek philosophy (cf. I 1, p. 12). On the other hand, it is only fairly recently that the potentialities of symbolic notation have been explored – probably through the inspiration of Leibniz, and systematically only since the 19th century. (For the historical aspects, the reader is referred to Scholz [1], Bocheński [1] and Kneale-Kneale [1]).

1. FROM TRADITIONAL LOGIC TO 'LOGISTIC'

We shall illustrate the three above-mentioned stages of abstraction in terms of a simple mode of inference taken from what is nowadays termed propositional logic. First, the method of typical examples:

(a) If Ringo has won at Bingo, Ringo becomes irresponsible. Ringo does not become irresponsible. *Therefore*, Ringo has not won at Bingo.

(b) It is has rained, it is wet. It is not wet. *Therefore*, it has not rained.

(c) If 99 is divisible by 32, then 99 is divisible by 2. 99 is not divisible by 2. *Therefore*, 99 is not divisible by 32.

Clearly, these inferences are all based on the following simple rule of inference:

(d) If the first (holds[1]), then the second (holds). The second (does) not (hold). *Therefore*, the first (does) not (hold).

'If the first, then the second' and 'not the second' are called the *premises* (assumptions) and 'not the first' the *conclusion* of the rule of inference.

A whole system of propositions, or inferences, of this kind was formulated in roughly this form in the so-called Stoic logic (cf. Mates [1]). Before going on to formulate our example in terms of a symbolic notation, we shall compare (d) with the examples given under (a)–(c). Whoever asks himself whether (d) is a correct mode of inference, will answer this question in the affirmative, and will therewith accept also (a), (b) and (c).

However, examples (a) and (b), at any rate, may well evoke a response of 'Yes, but ...' from some people. This may be connected with the fact that in ordinary linguistic usage the precise meaning of a statement is usually determined only by the situational context. Only in this way do the contents of different parts of a proposition correspond so precisely with one another that they can be referred to as 'the first' and 'the second' etc. Thus, in example (a), the sentence 'Ringo does not become irresponsible' may be referred to as 'not the second' only if it is understood as a report about Ringo's behaviour and not in the sense of 'It is not Ringo's nature to become irresponsible.' On the other hand, if the first premise is interpreted in this latter sense (viz as 'If Ringo wins at Bingo, then it is Ringo's nature to become irresponsible'), then it cannot be regarded as a particular instance of the general rule that 'whoever wins at Bingo, becomes irresponsible'.

Whereas objections might be raised in the case of example (a) because of a certain qualitative ambiguity of the statements occurring in it, example (b), raises difficulties of a rather different kind. Here we will be inclined to doubt the applicability of the general rule of inference (d) because of a quantitative indeterminateness of the concepts used: How many drops constitute a shower? What length of time counts as a time after a shower?

With example (c) the situation is quite different. The concepts occurring in it all belong to that branch of knowledge that has most effectively freed itself from the indeterminateness of ordinary linguistic usage.[2] Not only those who know what 'divisible' means in mathematics, but everybody who knows that it is a well-defined concept will admit (c) to be an instance of (d).

However, a rule of inference such as (d) need not be restricted in its application to those branches of knowledge that have clear-cut concepts. A sufficiently close correspondence in meaning between the parts of statements in given premises, may also be based on particular experience and knowledge. The situation is similar to that of the application of geometry (the theory of space) as a pure theory to the 'world we live in'. Here there are no perfect points, straight lines, planes, spheres etc.; nevertheless, those who have the requisite specialized knowledge know in precisely what sense and with what degree of accuracy the laws of geometry may be applied to triangulation points, straight roads, walls, balloons and so forth. But they are also aware that it would be wrong to try to base geometry on such partial approximations as, for example, walls – instead of on perfect, ideal planes. A general law is more readily grasped if questions concerning e.g. the difference between a real wall and an ideal plane, important as they are in practice, are temporarily left out of account, since they merely confuse the issue.

In logic, similarly, we must take account on the one hand of the pure forms of inference and on the other, of the linguistic representations of the formal structures occurring in these forms. Indeed, in most cases we arrive at the formal structures only *via* a linguistic representation; for we must be able to talk about them. However, instead of 'talking' about them we can use a symbolic notation, and this more direct way is today much used to describe such structures.

The development of logic as a science no doubt started with the collection of examples of particular modes of inference. But even in classical Greek philosophy we find two subdivisions of logic being investigated and systematized in form (2). These are in the first place the so-called

syllogistic of Aristotle (cf. I 2 and 3), and secondly the **propositional logic** of the Stoic school (Chrysippos and others), which has already been mentioned in connexion with example (d).

On the other hand, neither ancient nor scholastic logic appears to contain a systematic formulation of the **logic of relations**, although Greek geometry[3] could not have been as highly developed as it was without a rational treatment of relations such as (the point P) *lies on* (the straight line s) or *the distance between* (the two points AB) *is the same as* (that between the two points CD). But the logic that is applied in Greek geometry is not generally applied in a way that consciously points to the underlying forms of inference.

Among the deductions of the scholastics, too, there are a number of inferences that nowadays are recognized as examples of the logic of relations.[4]

Modern logic – developed from the 19th century onwards in the work of Boole, Schröder, Peirce, Frege, Peano, Whitehead, Russell and others – has produced a totality of proofs and modes of inference within which the work of Aristotle and the Stoics falls naturally into place, but which contains in addition a comprehensive theory of relations. Thus modern logic, which is often referred to as **logistic**, differs from traditional logic only in that it is much more inclusive. This development, however, was made possible only through the systematic use of **symbolic** techniques, i.e. symbolic notation, by means of which even complex meanings can be formulated in simple and significant terms. For this reason modern logic is often called **symbolic logic**.

The following example will show the increased perspicuity resulting from the introduction of a symbolic notation. The so-called binomial theorem can be formulated in words as follows:

> The square of the sum of two numbers equals the sum of the following summands: the square of the first number, twice the product of the two numbers, and the square of the second number.[5]

Compare this with the symbolic formulation of the theorem:

$$(a + b)^2 = a^2 + 2ab + b^2.$$[6]

Similarly, we can obtain significant formulations for logical laws through the introduction of suitable symbols. Using the notation that will be described in this book (cf. III 2, p. 50 f.; 3, p. 56 f.), the law expressed in (d) is formulated as follows:

(e) $A \rightarrow B, \quad \neg B \models \neg A,$

a suggested way of reading this being: 'if *A*, then *B*' and 'not *B*', then 'not *A*'. On the basis of this translation the reader might care to examine the following sentences and to ascertain the correctness of the inferences presented:

(f) $A \rightarrow \neg B, \quad B \models \neg A$
(g) $\neg A \rightarrow B, \quad \neg B \models A$
(h) $\neg A \rightarrow \neg B, \quad B \models A$

A systematic method for the construction of such 'rules' is set forth in IV 2, p. 70 f.

One branch of modern logic that scarcely occurs in traditional logic is that concerned with the investigation of the possibilities and limits of the symbolic or formal method in logic. These enquiries into logic are also known as metalogic. They make up an ever-increasing proportion of the work done in symbolic logic. Some typical, metalogical problems are put forward in Sections VI and VII.

2. CONVENTIONS REGARDING BORDER-LINE CASES

Modern logic has been developed to a large extent by mathematicians. They have transferred to logic the well-tried method of mathematics of simplifying complex conceptual structures by appropriate conventions about border-line cases. (Thus the decimal notation, familiar to all of us, was made possible only by the introduction of zero as a number or cipher.) Because of this, the presentation of some parts of traditional logic within the framework of modern logic differs from that of traditional logic, but without essential divergence of meaning. We shall illustrate this from **Aristotelian syllogistic** (including, where relevant, traditional logic, which is based on Aristotelian logic and in part diverges from it).

In Aristotelian syllogistic logical propositions or modes of inference of the following form are put forward:

(a) If *A* and *B*, then *C;*

or alternatively[7]

(b) From *A* and *B* follows *C;*

or alternatively[7]

(c) *A*, *B;* therefore *C;*

where *A*, *B*, *C* are propositions having one of the four following forms:

(1) All things having the property *S* have the property *P*, or briefly: all *S*-things are *P*-things, or: all *S* are *P;* traditionally symbolized by the formula *S***a***P*.

(2) Some *S*-things are *P*-things, or: at least one *S*-thing is a *P*-thing, traditionally symbolized by *S***i***P*.

(3) All *S*-things are *not P*-things, or: no *S*-thing is a *P*-thing, traditionally symbolized by *S***e***P*.

(4) Some *S*-things are *not* *P*-things, traditionally symbolized by *S***o***P*.[8]

Importance attaches only to the relation to be expressed between the so-called *subject-term*[9] *S* and the *predicate-term*[9] *P*, and this relation may be expressed in yet other ways than indicated above. Thus in general the form *S***a***P* will include propositions such as 'Relatives are human beings' and 'The best things in life are free', whereas in a sentence like 'The last mile is the most difficult' the definite article in 'the last mile' may be taken either in the sense of 'every' or 'this', depending on the context. This aspect will be further discussed in III 1, p. 48.

A relation between *A* and *B*, the *premises*, and *C*, the *conclusion* of (a), (b) or (c) is set up in accordance with the following rule:

The subject-term of *C* is to appear in *B*, and the predicate-term of *C* in *A*, either as subject or as predicate; and *A* and *B* are to have a term, the so-called middle-term in common.[10] From among the propositions of form (a) or (b) fulfilling these conditions – of which there are no fewer than 256 – those are to be selected that express logical propositions, correct inferences or 'admissible' rules.

The positions of the subject-term *S* and the predicate-term *P* in the premises *A*, *B* and therewith the position of the middle-term *M* in *A* and *B* having been decided – this determines the so-called figure of the inference – attention must also be paid to the connectives **a**, **i**, **e**, **o** holding between the terms *M*, *P;* *S*, *M;* *S*, *P*.

Traditional logic puts forward 19 propositions or modes of inference that satisfy the above conditions and are not weak forms of other correct inferences (of these latter there are five so-called subaltern modes of inference.) We shall arrange them, as is customary, according to their 'figures', and shall write '$\frac{\begin{matrix}A\\B\end{matrix}}{C}$' as abbreviations for (a) resp. (b), (c). Occasionally we shall employ the equivalent forms '$\frac{A\ B}{C}$' or '$\frac{B\ A}{C}$' as being more appropriate.

Figure 1 comprises four modes of inference

$$\frac{\begin{matrix}M\mathbf{a}P\\S\mathbf{a}M\end{matrix}}{S\mathbf{a}P}\ (1.1)\qquad \frac{\begin{matrix}M\mathbf{e}P\\S\mathbf{a}M\end{matrix}}{S\mathbf{e}P}\ (1.2)\qquad \frac{\begin{matrix}M\mathbf{a}P\\S\mathbf{i}M\end{matrix}}{S\mathbf{i}P}\ (1.3)\qquad \frac{\begin{matrix}M\mathbf{e}P\\S\mathbf{i}M\end{matrix}}{S\mathbf{o}P}\ (1.4)$$

Figure 2 also comprises four modes of inference

$$\frac{\begin{matrix}P\mathbf{e}M\\S\mathbf{a}M\end{matrix}}{S\mathbf{e}P}\ (2.1)\qquad \frac{\begin{matrix}P\mathbf{a}M\\S\mathbf{e}M\end{matrix}}{S\mathbf{e}P}\ (2.2)\qquad \frac{\begin{matrix}P\mathbf{e}M\\S\mathbf{i}M\end{matrix}}{S\mathbf{o}P}\ (2.3)\qquad \frac{\begin{matrix}P\mathbf{a}M\\S\mathbf{o}M\end{matrix}}{S\mathbf{o}P}\ (2.4)$$

Figure 3 consists of six modes of inference, the last two of which (marked *) provide examples on the topic 'conventions about border-line cases'.

$$\frac{\begin{matrix}M\mathbf{a}P\\M\mathbf{i}S\end{matrix}}{S\mathbf{i}P}\ (3.1)\qquad \frac{\begin{matrix}M\mathbf{e}P\\M\mathbf{i}S\end{matrix}}{S\mathbf{o}P}\ (3.2)\qquad \frac{\begin{matrix}M\mathbf{i}P\\M\mathbf{a}S\end{matrix}}{S\mathbf{i}P}\ (3.3)\qquad \frac{\begin{matrix}M\mathbf{o}P\\M\mathbf{a}S\end{matrix}}{S\mathbf{o}P}\ (3.4)$$

$$\frac{\begin{matrix}M\mathbf{a}P\\M\mathbf{a}S\end{matrix}}{S\mathbf{i}P}\ (3.5)^{*}\qquad \frac{\begin{matrix}M\mathbf{e}P\\M\mathbf{a}S\end{matrix}}{S\mathbf{o}P}\ (3.6)^{*}$$

Figure 4 includes five modes of inference, the last two of which again furnish examples on the subject 'conventions about border-line cases'.

$$\begin{array}{c} P\mathbf{a}M \\ M\mathbf{e}S \\ \hline S\mathbf{e}P \end{array} \quad (4.1) \qquad \begin{array}{c} P\mathbf{e}M \\ M\mathbf{i}S \\ \hline S\mathbf{o}P \end{array} \quad (4.2) \qquad \begin{array}{c} P\mathbf{i}M \\ M\mathbf{a}S \\ \hline S\mathbf{i}P \end{array} \quad (4.3)$$

$$\begin{array}{c} P\mathbf{a}M \\ M\mathbf{a}S \\ \hline S\mathbf{i}P \end{array} \quad (4.4)^* \qquad \begin{array}{c} P\mathbf{e}M \\ M\mathbf{a}S \\ \hline S\mathbf{o}P \end{array} \quad (4.5)^*$$

As there is no clear-cut systematic connexion between these 19 modes of inference, Latin mnemonics were later introduced for them. For example, in Figure 1, (1.1) = *barbara*, (1.2) = *celarent*, (1.3) = *darii*, (1.4) = *ferio*, the consonants chosen expressing relations to modes of inference belonging to other figures. In former times these mnemonics were committed to memory much as one might learn a poem.

The reader is recommended to write out some of these inferences in words, and to convince himself of their validity. Our discussion here will be limited to the four inferences marked *, as their validity depends on a special convention, which concerns propositions of the form *S***a***P* and *S***e***P*. In ordinary language a proposition of the form 'All *S*-things are *P*-things' is *immediately* meaningful only if there are *S*-things (if the meaning of the term *S* is *non-empty*), and is in this case equivalent to

(d) For *all things x:* if *x* is an *S*-thing, then *x* is a *P*-thing.

Whereas Aristotle and his successors interpreted the proposition *S***a***P* as also asserting the proposition that there are *S*-things, it has since proved more convenient to agree to regard *S***a***P* as equivalent in meaning to (d) in all cases. This convention does not affect the 'natural-language' use of *S***a***P*. On the other hand, it removes the need to make special case-distinctions for inferences that exceed the limits set by (a) or (b) but are nevertheless of practical importance. And in any case, which logic could we apply if we first had to show that a newly defined term was non-empty?

The logical schema (of which we shall not here give an explicit formulation) underlying the inference

(e) From the fact that all sailors are men, it follows that all good sailors are good men.[11]

should hold irrespective of whether there are good sailors or not. At the same time the modern convention makes it easier to integrate the traditional schemata – excepting, of course, those marked * – within a more inclusive system.

Similarly *S***e***P* is to be understood as equivalent to

(f) For *all things x:* if *x* is an *S*-thing, then *x* is not a *P*-thing.

The schemata marked * depend on the traditional convention for the following reasons. In the schemata

$$\frac{M\mathbf{a}P \quad M\mathbf{a}S}{S\mathbf{i}P}\ (3.5) \qquad \frac{M\mathbf{e}P \quad M\mathbf{a}S}{S\mathbf{o}P}\ (3.6) \qquad \frac{P\mathbf{e}M \quad M\mathbf{a}S}{S\mathbf{o}P}\ (4.5)$$

the conclusion *S***i***P*, resp. *S***o***P*, viz 'Some *S*-things are (not) *P*-things', in all cases presupposes 'namely the *M*-things'. (The reader should try to find examples for *S*, *M*, *P* so that – under the modern convention – the premises are true and the conclusions false.) On the other hand, the schema

$$\frac{P\mathbf{a}M \quad M\mathbf{a}S}{S\mathbf{i}P}\ (4.4)$$

is based – less obviously perhaps – on the fact that in the traditional interpretation the premise *P***a***M* implies the existence of *P*-things. A counter-example on the basis of the modern convention would be: Every lunar man is (at any rate) a man; every man is a living being; and falsely: some living beings are lunar men.

3. THE SYMBOLIC METHOD – ILLUSTRATED IN TERMS OF SYLLOGISTIC

Even in traditional logic we find a symbolic notation used to represent the syllogisms. However, the full advantages of symbolism become apparent only when this is used not merely for writing out the modes of inference, but also for establishing their foundation. Although on the whole the preference nowadays is to integrate Aristotelian syllogistic within a more inclusive system of modes of inference (cf. IV 2, p. 78), in the sense of the 'translations' in I 2, (d) and (f), it is also possible – and highly instructive – to apply symbolic method to the task of systematizing the Aristotelian schemata themselves, so as to bring out clearly the interconnexions existing between them, although these were, of course, also known to traditional logic.

With the introduction of three auxiliary modes of inference all 19 modes of inference can be reduced to the schemata *barbara* and *darii*, i.e. (1.1) and (1.3). More far-reaching reductions can be obtained by an increased use of symbolic notation, as will be indicated below.

The reduction to *barbara* and *darii* may be represented in the following stages:

(1) Reduction of the connectives **e**, **o** to **a**, **i** by the introduction of complementary terms. A proposition of the form 'All *S*-things are not *P*-things' may be interpreted on the one hand as 'All *S*-things are-not *P*-things' and on the other hand as 'All *S*-things are non-*P*-things', i.e. in the one case the property *P* is *denied* in respect of *S*-things, and in the other the property non-*P* is *affirmed* in respect of *S*-things. We call non-*P* the *complementary term to P*.[12] It is symbolized by '**n***P*'. Such formulation in terms of '**n***P*' is, of course, equivalent to the original one, but proves to be useful in that it allows a mode of inference formulated for any term whatsoever to be applied also to complements of terms.

We can now define '*S***e***P*' by '*S***an***P*' and '*S***o***P*' by '*S***in***P*', or briefly:

$$\mathbf{e} =_{\mathrm{Df}} \mathbf{an} \qquad \text{(to be read as '}\mathbf{e}\text{ equals }\mathbf{an}\text{ by definition')}$$
$$\mathbf{o} =_{\mathrm{Df}} \mathbf{in}.$$

However, this procedure needs to be justified. On substituting **n***P* for *P*, **n** and *P*, of course, belong together. This can be expressed by brackets: '*S***a**(**n***P*)'. We then introduce a new concept (**an**) by means of $S(\mathbf{an})P =_{\mathrm{Df}} S\mathbf{a}(\mathbf{n}P)$. Since on the basis of the definition both bracketings are equivalent, it is agreed to omit them, so that '*S***an***P*' and '*S***e***P*' are interchangeable. We argue similarly concerning '**o**' and '**in**'.

It should be noted that these reductions depend on the choice of a suitable symbolism. For example, if instead of '*S***a***P*' we had written '**a***SP*', corresponding simplifications would have been possible only for the complements of subject-terms, which are less important in this connexion.[13]

(2) Implications of the admission of complementary terms. Since we are concerned, in all the modes of inference investigated here, only

with the extensions of terms (the extension of a term being the totality of things subsumed by it) and since the complement of the complement of a term *P* is co-extensional with *P* – is, indeed, identical with *P* if terms are regarded as identical with their extensions – any term *P* in any mode of inference may be replaced by **nn***P* as required; examples will be given under (4).

Complementary terms having once been admitted, there is no reason why they should not also occur as subjects in propositions, i.e. why propositions of the form **n***S***a***P*, **n***S***an***P*, **n***S***i***P*, **n***S***in***P* should not be accepted. In this way some of the proposed *derivations* are simplified. On the other hand, the task set in I 2, p. 14 f. is enlarged, if propositions having the new forms are accepted for *A*, *B*, *C*. This extended form of syllogistic was, in fact, investigated by traditional logic, but did not become established (cf. in this connexion Reichenbach [1], and II 3, p. 44).

(3) The basic modes of inference. The decision to which of the Aristotelian modes of inference the remainder are to be reduced, is to some extent an arbitrary one. However, if we select **a**, **i**, **n** as 'basic connectives', then the following choice recommends itself as being a 'natural' one. We select as 'basic':

(a) The modes of inference *barbara* and *darii*, which we quote again in a form suitable in this connexion:

$$\frac{S\mathbf{a}M \quad M\mathbf{a}P}{S\mathbf{a}P} \quad (1.1) \qquad \frac{S\mathbf{i}M \quad M\mathbf{a}P}{S\mathbf{i}P} \quad (1.3)$$

(b) Two auxiliary modes of inference, which express symmetry properties of **i** and **an** (known in traditional logic as *conversio simplex* or 'simple conversion'):

$$\frac{S\mathbf{i}P}{P\mathbf{i}S}, \qquad \frac{S\mathbf{an}P}{P\mathbf{an}S}$$

(c) Two auxiliary modes of inference of a more general nature, which express the substitutability formulated in (2). (Let ... *P* ... be a proposition in which *P* occurs.)

$$\frac{\ldots P \ldots}{\ldots \mathbf{nn}P \ldots} \qquad \frac{\ldots \mathbf{nn}P \ldots}{\ldots P \ldots}$$

(d) An auxiliary mode of inference that depends for its validity on the subject-term being non-empty, and which thus expresses this presupposition. (Thus the modes of inference 'derived' with its aid depend on the traditional convention discussed in I 2, p. 16 f.)

$$\frac{S\mathbf{a}P}{S\mathbf{i}P}\;(*)$$

(The asterisk is intended as a reminder of the special role of this inference.) If we had formulated the logical laws contained in (a), (b), (c), (d) as propositional forms, instead of as modes of inference, then these propositions would be designated as *axioms*. But in this case special modes of inference (belonging to propositional logic) would be required to express deductions from the axioms. On the other hand, the inter-connexions between modes of inference can be very clearly symbolized, as will be shown below.

(4) Inter-connecting two modes of inference is possible only by using the conclusion of the one as (some) premise of the other. This will be symbolized by the immediate juxtaposition of the inferences or[14] modes of inference, as the following examples show:

$$\dfrac{\dfrac{S\mathbf{a}P}{S\mathbf{i}P}\;(*)}{P\mathbf{i}S} \qquad \dfrac{\dfrac{S\mathbf{a}P}{S\mathbf{ann}P}}{\mathbf{n}P\mathbf{an}S} \qquad \dfrac{\dfrac{\mathbf{n}P\mathbf{an}S}{S\mathbf{ann}P}}{S\mathbf{a}P}$$

In this way derived modes of inference are obtained, the uppermost formulas being the premises and the bottom one the conclusion. (The last two examples are also known as contrapositions. Note the complementary terms in subject place.)

If at least one mode of inference with two premises is used, we obtain a tree-like figure. In this case the chain of inferences to be worked through is not pre-determined, since we can choose which of the two branches of the tree meeting in an inference we wish to work out first. However, we can also proceed differently and begin by constructing the whole tree in its final form, either mentally or on paper, and then test the separate inferences in any order of sequence, to see whether they follow from a basic mode of inference.

This way of representing derivations or proofs in the form of a tree is much used in modern logic, as it allows the assumptions made at every stage of a proof to be clearly indicated. However, it has the disadvantage that if the same assumption is used several times, the derivation must be written out in full on each occasion. A way of avoiding this will be shown in IV 2, p. 72.

(5) The derived modes of inference. We give below the 'trees' for the derived inferences, arranged according to the basic modes of inference used. In this way the derived modes are in a sense analysed out into the basic ones.

(a) Inferences reducible to (1.1) and simple conversions:

$$\frac{S\mathbf{a}M \quad M\mathbf{an}P}{S\mathbf{an}P} \qquad \text{i.e. (1.2)}$$

Here $\mathbf{n}P$ has been substituted for P; this is indicated by '$P/\mathbf{n}P$'.

$$\frac{S\mathbf{a}M \quad \dfrac{P\mathbf{an}M}{M\mathbf{an}P}}{S\mathbf{an}P} \quad \text{i.e. (2.1)}$$

$$\frac{\dfrac{P\mathbf{a}M \quad M\mathbf{an}S}{P\mathbf{an}S}}{S\mathbf{an}P} \qquad \text{i.e. (4.1)}$$

$$\frac{\dfrac{P\mathbf{a}M \quad \dfrac{S\mathbf{an}M}{M\mathbf{an}S}}{P\mathbf{an}S}}{S\mathbf{an}P} \quad \text{i.e. (2.2)}$$

Through the use of simple conversions we obtain inferences that belong to other figures. The derived modes of inference are arranged according to the number of simple conversions required.

(b) An inference that, by way of a corollary, presupposes the non-emptiness of a term.

$$\frac{\dfrac{\dfrac{P\mathbf{a}M \quad M\mathbf{a}S}{P\mathbf{a}S}}{P\mathbf{i}S}\,(*)}{S\mathbf{i}P} \qquad \text{i.e. (4.4)}$$

The weakening of the premise involved in the inference marked (*) is required in order to obtain the form $S \ldots P$ prescribed for the conclusion in days gone by.

(c) Inferences based on (1.3) and simple conversions.

$$\frac{S\mathbf{i}M \quad M\mathbf{an}P}{S\mathbf{in}P} \quad \text{i.e. (1.4)} \qquad\qquad \frac{S\mathbf{i}M \quad \frac{P\mathbf{an}M}{M\mathbf{an}P}}{S\mathbf{in}P} \quad \text{i.e. (2.3)}$$

$$\frac{\frac{M\mathbf{i}S}{S\mathbf{i}M} \quad M\mathbf{a}P}{S\mathbf{i}P} \quad \text{i.e. (3.1)} \qquad\qquad \frac{\frac{M\mathbf{i}S}{S\mathbf{i}M} \quad M\mathbf{an}P}{S\mathbf{in}P} \quad \text{i.e. (3.2)}$$

$$\frac{\frac{P\mathbf{i}M \quad M\mathbf{a}S}{P\mathbf{i}S}}{S\mathbf{i}P} \quad \text{i.e. (4.3)} \qquad\qquad \frac{\frac{\frac{M\mathbf{i}P}{P\mathbf{i}M} \quad M\mathbf{a}S}{P\mathbf{i}S}}{S\mathbf{i}P} \quad \text{i.e. (3.3)}$$

$$\frac{\frac{M\mathbf{i}S}{S\mathbf{i}M} \quad \frac{P\mathbf{an}M}{M\mathbf{an}P}}{S\mathbf{in}P} \quad \text{i.e. (4.2)}$$

$$\frac{S\mathbf{in}M \quad \frac{\frac{P\mathbf{a}M}{P\mathbf{ann}M}}{\mathbf{n}M\mathbf{an}P}}{S\mathbf{in}P} \quad \text{i.e. (2.4)}$$

Here derivation from (1.3) requires the substitution of complementary subject terms.

$$\frac{\frac{\frac{M\mathbf{in}P}{\mathbf{n}P\mathbf{i}M} \quad M\mathbf{a}S}{\mathbf{n}P\mathbf{i}S}}{S\mathbf{in}P} \quad \text{i.e. (3.4)}$$

The use of complementary subject terms could be formally avoided by e.g. obtaining (3.4) from the previously derived (3.3) by the substitution $P/\mathbf{n}P$. But if this substitution is made in the derivation of (3.3), complementary subject terms result once again.

(d) Inferences where apart from (1.3) and simple conversion, the non-emptiness of a term is presupposed.

$$\dfrac{\dfrac{\dfrac{M\mathbf{a}S}{M\mathbf{i}S}(*)}{S\mathbf{i}M}\quad M\mathbf{a}P}{S\mathbf{i}P}\qquad \text{i.e. (3.5)}$$

$$\dfrac{\dfrac{\dfrac{M\mathbf{a}S}{M\mathbf{i}S}(*)}{S\mathbf{i}M}\quad M\mathbf{an}P}{S\mathbf{in}P}\qquad \text{i.e. (3.6) (derived from the preceding by } P/\mathbf{n}P.)$$

$$\dfrac{\dfrac{\dfrac{M\mathbf{a}S}{M\mathbf{i}S}(*)}{S\mathbf{i}M}\quad \dfrac{P\mathbf{an}M}{M\mathbf{an}P}}{S\mathbf{in}P}\qquad \text{i.e. (4.5)}$$

(e) With regard to those modes of inference whose mere formulation requires complementary subject terms: they, too, can all be reduced to (1.1) and (1.3) by substitutions and simple conversions. We give only one example, as a complete enumeration would lead us too far from our main topic.

$$\dfrac{\mathbf{n}S\mathbf{a}M\quad M\mathbf{a}P}{\mathbf{n}S\mathbf{a}P}$$

(6) More far-reaching reductions are obtained on the basis of the following considerations:

(a) Every proposition of the form

not all S-things are not P-things

is equivalent to

some S-things are P-things.

(b) Every proposition of the form

If *A* and *B*, then *C*

is equivalent to

If *A* and not *C*, then not *B*

resp. If *B* and not *C*, then not *A*.

Before giving a formal derivation we give below a proof formulated in a 'mixed style', using '$\neg$' for 'not' (where this denies a proposition[15]). Substituting *P*/**n***P* in (1.1), we obtain (1.2)

$$\frac{S\mathbf{a}M \quad M\mathbf{an}P}{S\mathbf{an}P}$$

This is equivalent, on the basis of (b), to

$$\frac{S\mathbf{a}M \;\neg\; S\mathbf{an}P}{\neg\; M\mathbf{an}P}$$

On the basis of (a), this in turn is equivalent to

$$\frac{S\mathbf{a}M \quad S\mathbf{i}P}{M\mathbf{i}P} \quad \text{and to} \quad \frac{S\mathbf{i}P \quad S\mathbf{a}M}{M\mathbf{i}P}$$

Since no significance attaches to the choice of the letters *S*, *M*, *P* they can be interchanged by the simultaneous substitutions *S*/*M*, *M*/*S*, *P*/*S*.[16] This clearly yields the mode of inference below (apart from the interchange of *S* and *P*)

$$\frac{M\mathbf{i}S \quad M\mathbf{a}P}{P\mathbf{i}S} \quad (3.3)$$

and by means of two simple conversions

$$\cfrac{\cfrac{S\mathbf{i}M}{M\mathbf{i}S} \quad M\mathbf{a}P}{\cfrac{P\mathbf{i}S}{S\mathbf{i}P}}$$

Within the framework of the basic modes of inference stated under (3), (1.3) can thus be replaced by more basic inferences, if the transformations on the basis of (a) and (b) are put into symbolic form. This, however, is not a question of finding an equivalent replacement, but of integration within a richer system, as e.g. the symbol $\neg$ has not been used previously.

It would be easy to integrate (a) within the system stated by (3) and (4), viz by introducing the following modes of inference:

$$\frac{\neg\, S\mathbf{an}P}{S\mathbf{i}P}, \qquad \frac{S\mathbf{i}P}{\neg\, S\mathbf{an}P},$$

and where appropriate also

$$\frac{S\mathbf{an}P}{\neg\, S\mathbf{i}P}, \qquad \frac{\neg\, S\mathbf{i}P}{S\mathbf{an}P}.$$

On the other hand, the formal representation of (b) is less easy, since here the transition from $\frac{A \;\; B}{C}$ to $\frac{A \;\neg\, C}{\neg\, B}$ and *vice versa* would have to be reduced essentially to a linking of modes of inference. However, the effort involved is compensated by the fact that the result – after some obvious additions – is the whole of propositional logic. Cf. in this connexion IV 2, p. 73 f. We cannot deal with this topic more fully here, and shall merely state that apart from linking the modes of inference we should also require a principle for the elimination of premises, since the transitions obviously involve the elimination of the premises B, resp. $\neg C$, as such.

The modes of inference with which we have so far concerned ourselves may be further reduced by being integrated within so-called predicate logic. For this purpose the translations given in I 2, p. 16 f. of $S\mathbf{a}P$, $S\mathbf{e}P$ etc. [cf. *ibid.* (d), (f)], are expressed symbolically.

Anticipating the symbolism used in III 2 A, p. 51 and C, p. 52 f, we have the following definitions:

$$S\mathbf{a}P =_{\mathrm{Df}} \bigwedge x(Sx \rightarrow Px),$$
$$S\mathbf{e}P =_{\mathrm{Df}} \bigwedge x(Sx \rightarrow \neg\, Px),$$
$$S\mathbf{i}P =_{\mathrm{Df}} \bigvee x(Sx \wedge Px),$$
$$S\mathbf{o}P =_{\mathrm{Df}} \bigvee x(Sx \wedge \neg\, Px).$$

Although lack of space does not permit a fuller exposition, we should like at least to mention that in this way the above discussed modes of inference are reduced to the rules of inference that result from the meanings of the symbols on the right-hand side of the equivalences (cf. IV 2, p. 77).

NOTES

[1] The expression 'the first' really stands for a proposition; 'holds' and 'does hold' are mere concessions to grammatical usage.

[2] As we shall be concerned in this book to criticize such indeterminateness, we should like to draw attention here to its value and use: it is precisely this quality, in conjunction with our ability to understand incomplete statements, that allows natural languages to adapt themselves to entirely new situations.

[3] In the form of Euclid's Elements.

[4] J. Bendiek [1] has examined some of these inferences with the aid of modern logical techniques.

[5] The 'square of a number' is a term taken from geometry and is used to designate the product of a number with itself, in symbols: $a^2 = a \cdot a$.

[6] This formulation does not take account of the fact that the equivalence is asserted for any numbers *a*, *b* whatsoever. This omission is made good in II 2, p. 40 and III 2, p. 53 f.

[7] Cf. III 3, p. 61 and IV 1, p. 67 f. for a discussion of the difference between these formulations.

[8] This systematic formulation is not Aristotle's but derives from the time of traditional logic.

[9] Here 'term' refers sometimes to the symbol (as it stands 'terminally') but more often to the concept indicated. This ambiguity is no drawback here.

[10] A violation – usually a hidden one – of this condition is called a *quaternio (terminorum)*.

[11] The reader should consider this proposition first as an example of a correct form of inference and then as an example of a trick effect often encountered in logic: a '*so-and-so S*-thing' is in ordinary language often different from an *S*-thing that happens to have the property *so-and-so*.

[12] On problems connected with complementary terms, cf. II 3, p. 44.

[13] This may be merely a historical coincidence, due to the structure of language: on the other hand, it is possible that language developed in this way 'in order to' avoid, or at least evade, the possibility contained in '**n***P***a***P*' of making statements about all things in the world.

[14] Inference should be understood as the resultant of a complete instantiation (specification) of the variable terms used in the modes; but mixed cases are also possible.

[15] Cf. V 2 for a discussion of the difference between the negation of a proposition and the formation of the complement of a term; the two are, of course, connected.

[16] If these substitutions were carried out consecutively, the desired result would obviously not be achieved. It could, however, be obtained by the consecutive substitutions S/Q, P/S, M/P, Q/M.

CHAPTER II

LOGIC AS ONTOLOGY

If we wish to assert that something is quite certain, we often say that it is 'logical'. Although in fact examples of such use rarely fall within the sphere of logic, they nevertheless indicate a strong faith in logic. There is even a brand of cigarettes that has been advertised as the 'Logical Move'. And indeed it is much easier to doubt assertions such as 'Brutus murdered Caesar' or 'I was in Leicester on September 9th 1965', than for example the proposition:

(a) 'Every apple is sweet or there is at least one apple that is not sweet'

or any proposition having one of the forms discussed in I 2; 3.

To return to example (a): clearly, our faith in the truth of such propositions stems from the fact that they are not contingent on our particular experiences of apples etc. We might, after all, make similar statements about pears, or plums or potatoes, or replace the predicate 'is sweet' by 'is sour' or 'is yellow' etc. – if the reader still doubts the truth of proposition (a) he should check that he has not understood it in the sense of

(b) '*One knows* that every apple is sweet or *one knows* at least one apple that is not sweet.'

This or similar interpretations of (a) are put forward by a well-known school of logic, the so-called Intuitionist School. In this book, however, we wish to represent logic as a kind of theory of the general form of the 'world' – not as a theory of our knowledge of the 'world', which necessarily varies with time.

1. THE WORLD AS DOMAIN OF OBJECTS WITH PROPERTIES AND RELATIONS

It is a typical feature of propositions of the type of (a) above, that when we have grasped the truth of *one* proposition we often realize that this truth does not depend on the special meanings of certain words that occur in it (here 'apple', 'is sweet'[1]), but that every proposition of the form indicated by (a), viz:

(A) Every *A*-thing is a *B*-thing or there is some *A*-thing that is not a *B*-thing, is true.

A propositional schema that subsumes only true propositions is said to be generally valid.[2] Thus (A) is generally valid.

However, there are also schemata whose general validity is less readily grasped and requires special efforts. Logic is not concerned with personal convictions regarding the general validity of certain schemata but with this general validity itself and, where applicable, with the objective methods whereby the universal validity of one schema is derived from that of more basic schema. (This, of course, affords opportunities for establishing personal convictions.)

The following questions now arise:

(a) How is it that there are universally valid schemata?

(b) How can we grasp the universal validity of a schema, i.e. make judgments that 'exceed the bounds of all possible experience'?

(c) What insight regarding the 'real world' is afforded by such judgments?

The way in which we answer these questions depends, of course, on our philosophical standpoint. For example, (a) might be answered in the following different ways:

(1) From a realist point of view: The world of concrete (or abstract) things *consists of things*, which have some properties and not others, and between which some relations hold and not others. ('How do we know this?') It is this structure of the world and not the 'essence' of the things, properties and relations occurring in the world, that is relevant for the establishment of universal validity.

Let us call this 'picture' of the world 'discrete ontology',[3] its domain, the 'universe of discourse'.

(2) From an idealist point of view: The world appears to us the way we with our discrete ontology describe it, but the question whether it is *really* the way we describe it, is unanswerable or beside the point.

(3) From a 'fictionalist' point of view: We find it convenient to describe the world with a discrete ontology although its reality does not, or could not, fit our description.

(4) A philosophical standpoint of extreme scepticism will not provide any foundation for logic, and we shall therefore leave it out of account.

(5) Different degrees of selective scepticism are known which accept, for instance,

(a) only finite domains;

(b) infinite domains if these are regarded as being generated;

(c) actually infinite domains, but properties and relations only as being generated.

The methods here presented can be adapted to some of these philosophies.

Question (b) is a special case of the question: How is it possible to establish an ontology? Irrespective of whether one considers an ontology in the sense of a general theory of being as such, to be possible, this special question can, in our opinion, be answered along the following lines. The general validity of such schemata is based solely on the discrete ontology to which we made reference in answering question (a). The formal structure described by it may be investigated independently of how question (a) is answered. The interpretation and importance of this general validity, however, depends on the answer given to question (a).

The following considerations will provide an answer to question (c). For every generally valid schema of the forms

If A_1, then B; if A_1 and A_2, then B; etc.

there is (in a sense to be explained in IV 1, p. 67 f.) a mode of inference

$$\frac{A_1}{B}; \quad \frac{A_1\ A_2}{B}; \quad \text{etc.,}$$

where in every case the general validity depends on a relation between the form of A_1 and B; A_1, A_2 and B; etc., as e.g. in the case of the syllogisms (I 2, p. 14).

If the truth of a particular proposition of form A_1 (or of propositions of forms A_1, A_2) is not established by the form(s) of A_1 (and A_2 etc.) but by some experience, then by virtue of the mode of inference the truth of B is established on the basis of the same experience.

Let us take a simple example. A_1 expresses this experience; B is a different formulation of the same state of affairs. This, however, need not be immediately discernible but is inferred as indicated above.

Now take a more complicated case: Let A_1 express a totality of experiences such as is summarized or idealized[4] in a law of nature, and A_2 specific data of a blue-print of a machine, e.g. regarding speeds, gas pressures or electrical potentials, and let B describe the behaviour of the machine as inferred from A_1 and A_2. All that we need to do then is to spend time and money on the construction of the machines, whose previously inferred behaviour corresponds as far as possible to the intentions of their designer.

The modes of inference required by this example are usually counted as belonging to mathematics. In fact, however, mathematics in its modern form is a part of logic, namely that part which deals with numbers and spatial structures. This part of logic has for this reason been exhaustively developed by mathematicians. Greater emphasis used formerly to be placed on its application to numbers and spatial structures, whereas nowadays it is usually presented in a form where reference is made to things of any kind whatsoever having presupposed properties and relations. Mathematics in this form is distinguished from general logic only in that it deals preferably with such properties and relations as have proved themselves in the investigation of numbers and spatial structures.[5]

Apart from inessential variants, the discrete ontology referred to in our answer to question (a), determines the form of the language or conceptual notation[6] that we shall introduce here as the means of 'communicating' about this formal structure of the world. We symbolize as follows:[7]

(1) arbitrary objects, or 'individuals', i.e. things of any kind whatsoever by 'a', 'b', 'c' and also by 'x', 'y', 'z', adding distinguishing subscripts as required, e.g. 'a_0', 'a_1', 'a_2', ... The numerals have no counting function; their purpose is to make available as many symbols for things as are required. The use of different symbols (e.g. 'a' and 'c' or 'a_{17}' and 'a_{27}') is intended merely to express that the things symbolized thereby *can* be different;

(2.1) arbitrary properties by 'A^1', 'B^1', 'C^1', ..., with distinguishing subscripts as required, e.g. 'A_1^1', 'A_2^1', 'A_3^1', ...;

(2.2) arbitrary relations between two objects by 'A^2', 'B^2', 'C^2', also by 'A_1^2', 'A_2^2', 'A_3^2', ..., as required;

(2.3) arbitrary relations between three objects by 'A^3', 'B^3', 'C^3', and also by 'A_1^3', 'A_2^3', 'A_3^3', ..., as required;

(2.n) [By analogy relations between more than three (generally: n)

objects are symbolized[8] as required by 'A^n', 'B^n', 'C^n', or 'A_1^n', 'A_2^n', 'A_3^n', ...

(3.1) the fact that a property A^1 applies to an object a (i.e. that a has the property A^1), by 'A^1a';

(3.2) the fact that the relation A^2 holds between the things a and b, by 'A^2ab' (one occasionally finds the form 'aA^2b', which is modelled on the syntax of ordinary language, although the analogous form 'aA^1' has not established itself);

(3.3) the fact that the relation A^3 holds between the things a, b, c, by 'A^3abc';

(3.n) [By analogy '$A^na_1 \ldots a_n$' is introduced for any n whatsoever.]

Examples for (3.1), (3.2), (3.3), (3.n).

The fact that the property of being red belongs to a specific flower, may be expressed in ordinary language by the sentence: 'This flower is red.' If we were to apply convention (3.1) to ordinary language, we should have to say: 'is red this flower.' This 'standardized proposition' is obtained from 'A^1a' by instantiating the general symbols 'A^1' and 'a', viz by substituting the predicative linguistic component 'is red' for 'A^1' and the object-denoting component 'this flower' for 'a'. In general, however, it is advisable to adopt a natural-language syntax when substituting natural-language components for symbols in formulae.[9] Similarly we shall regard as a valid substitution in the formula 'A^2ab' not only the syntactically standardized proposition: 'the relation of being-taller-than[10] holds between James and Peter', but also the ordinary formulation of this state of affairs, viz: 'James is taller than Peter.' In an exactly parallel sense the assertion: 'Caesar's birth occurred between the founding of Rome and the migration of the Germanic tribes' will be regarded as a proposition of the form 'A^3abc' (and not e.g. 'aA^3bc').

2. RELATIONS BETWEEN CONCEPTS AND LOGICAL OPERATIONS

A world having the structure described by a discrete ontology would be exhaustively determined by a catalogue $\mathbf{C}_1$ – possibly infinite – of all cases where a property is realized (i.e. applies to an object), or a relation holds (i.e. between a pair of objects, a triplet of objects etc.). But even apart from the problematical assumption that there will be proper names available for all objects and for all the properties and relations involved, such a catalogue would be little help in finding one's way about in such a world. For in order to determine whether a certain object a does *not* possess a certain property A^1, one would have to look through the whole catalogue – and find an answer only after 'infinitely many steps', i.e. not at all. Let us consider an improved catalogue $\mathbf{C}_2$ having the following content.[11]

(1) Every case listed in $\mathbf{C}_1$ is listed also in $\mathbf{C}_2$.

(2.1) It holds for every property A^1 that is involved and every object a in the 'world', that if $\mathbf{C}_1$ does not state the applicability of A^1 to a, then $\mathbf{C}_2$ records the non-applicability of A^1 to a.

(2.2) It holds for every relation A^2 that is concerned and every pair of objects a, b that if $\mathbf{C}_1$ does *not* state that A^2 holds between a and b, then $\mathbf{C}_2$ records the fact that A^2 does *not* hold between a and b.

In general

(2.n) It holds for every relevant relation A^n and every n-tuple $a_1, \ldots, a_n$ of objects that if $\mathbf{C}_1$ does *not* state that A^n holds between $a_1, \ldots, a_{n-1}$ and a_n, then $\mathbf{C}_2$ records the fact A^n does *not* hold between $a_1, \ldots, a_{n-1}$ and a_n.

In the following we shall use this general formulation also for the border-line case where $n=1$, i.e. where properties are concerned instead of relations; and we shall refer to these properties and relations as concepts.

The general form of catalogue $\mathbf{C}_2$, as described by (1), (2.1), (2.2), ... can be simplified if the non-applicability of a predicate is represented symbolically. We therefore stipulate as follows: let '$\neg A$' be the predicate that expresses the non-applicability of 'A'. Hence $\mathbf{C}_2$ can be described as follows:

For every relevant concept A^n (i.e.: $n=1, 2, 3, \ldots$) and every n-tuple of objects $a_1, \ldots, a_n$, $\mathbf{C}_2$ contains exactly one of the two expressions $A^n a_1 \ldots a_n$ or $\neg A^n a_1 \ldots a_n$.[12] This follows from the above specifications for determining $\mathbf{C}_2$ from $\mathbf{C}_1$.

Let us suppose a part-catalogue $\mathbf{C}_{A^n}$, extracted from $\mathbf{C}_2$, consisting of exactly those expressions in which a specific concept A^n occurs. Thus $\mathbf{C}_{A^n}$ says everything about A^n that can be known about A^n, i.e. to which n-tuples A^n applies. Let us now express this somewhat differently: C_{A^n}

describes a so-called logical function, viz that function[13] which ascribes the value '*true*', symbolically: '**T**', to every n-tuple $a_1, \ldots, a_n$ characterized by the occurrence of $A^n a_1 \ldots a_n$ in catalogue $\mathbf{C}_2$, and the value '*false*', symbolically: '**F**', to every n-tuple $a_1, \ldots, a_n$ characterized by the occurrence of $\neg A^n a_1 \ldots a_n$ in $\mathbf{C}_2$. The logical function correlated in this way with the concept A^n, is also called the course-of-value of A^n. Indeed, it is often said that a concept is nothing other than a function in this sense. Logical functions are also called attributes, so as to indicate that they are specifications (or idealizations) of properties and relations. We shall adopt this linguistic usage.

So far then, an attribute is merely a counterpart to the catalogue $\mathbf{C}_{A^n}$ of a concept A^n, and is even more abstract than it. However, instead of thinking in terms of the applicability or non-applicability of the concept A^n, to the n-tuple $a_1 \ldots a_n$, we can now refer to operations with the two '*truth values*' **T** and **F**. In this connexion we shall write '$A^n(a_1, \ldots, a_n)$' for the correlated truth value. That is, $A^n(a_1, \ldots, a_n)$ is **T** if $A^n a_1 \ldots a_n$, and **F** if $\neg A^n a_1 \ldots a_n$.

In order to find our bearings in a world described by a catalogue $\mathbf{C}_2$, let us try to find relations between the concepts or attributes A^n whose catalogues $\mathbf{C}_{A^n}$ are contained in $\mathbf{C}_2$. In the simplest cases such relations are represented by propositions in which 'atomic propositions' occur, e.g.

(1) $A^1 a$ and $\neg B^1 a$,

(2) It is credible that $A^1 a$ and $\neg B^1 a$,

(3) There are objects a having the property which is expressed by $A^1 a$ and $\neg B^1 a$ (or briefly: For some objects a, $A^1 a$ and $\neg B^1 a$).

We shall disregard (3), as the symbol 'a' is evidently used in a different sense here than in examples (1) and (2); (3) is not a proposition about the object a. The difference between (1) and (2) is the following: whether it is true that $A^1 a$ and $\neg B^1 a$ depends only on whether $A^1 a$ is true or false and on whether $B^1 a$ is true or false; on the other hand, whether it is credible that $A^1 a$ and $\neg B^1 a$ depends rather on what we *know* about similar cases. For example, a may be a specific almond in a bag, A^1 the property of being an almond, B^1 the property of being bitter. Then the proposition 'this is an almond and it is not bitter' would represent (1) (in a form syntactically adapted to natural language), and this proposition might be true for almond a (a fact that one would have to ascertain).

On the other hand, the truth of (2), which expresses the credibility of (1), depends much less on the truth of (1) as on e.g. how the almonds so far taken from the bag have tasted.

It will be more convenient to deal first only with relations where complexities as in (2) are excluded. These relations are thus wholly, and the corresponding propositional compounds are essentially determined by the fact that for every value **T** or **F** of the atomic propositions occurring in them, the value of the compound proposition is fixed; i.e. the relation holds in precisely those cases where the value of the compound proposition is **T**.

The simplest compound propositions of this kind are to be found in natural languages, for these have developed among other things out of the need for communication about an environment whose structure we have attempted to describe, at least approximately, by introducing the notion of catalogues of form $\mathbf{C}_2$. (This aspect will be discussed more fully in III 2, p. 55. Here we are concerned with the relations themselves, and not with their linguistic representations.[14] We shall refer to these only occasionally, when this makes for a simpler formulation.)

Clearly, these relations are logical functions. For if the places of the occurrence of atomic propositions in compound propositions are numbered (places where the same atomic proposition occurs being allocated the same number), and if n numbers are used, then the relation is described by a function that correlates one truth-value to every n-tuple of truth-values. Since only truth-values or n-tuples of truth-values occur as arguments, it is customary to speak of truth-functions in this connexion: thus of monadic, dyadic and n-adic truth-functions,[15] where the arguments are truth-values, pairs of truth-values and n-tuples of truth-values, respectively.

Truth-functions can be represented by means of catalogues called 'truth-tables' or 'matrices', in much the same way as attributes, but more simply. These matrices are always finite, since in the case of an n-term truth-function they contain exactly 2^n entries (**T** and **F** yield 2^n n-tuples, i.e. 2 for $n=1$, 4 for $n=2$, 8 for $n=3$ etc.). Since for each entry there are the two possibilities **T** and **F**, we thus have 4 truth-functions for $n=1$ (viz 2^2), 16 truth-functions for $n=2$ (viz 2^4), 256 truth-functions for $n=3$ (viz 2^8), and more than 8000 truth-functions for $n=4$.[16]

We give below the matrices for the four monadic truth-functions, where

π stands for the argument, $\phi_1(\pi)$ for the value of the first monadic function, $\phi_2(\pi)$ for that of the second, etc.

π	$\phi_1(\pi)$	$\phi_2(\pi)$	$\phi_3(\pi)$	$\phi_4(\pi)$
T	**T**	**T**	**F**	**F**
F	**T**	**F**	**T**	**F**

ϕ_1 and ϕ_4 are obviously trivial, since they are in no sense dependent on the value of π. ϕ_2 is likewise trivial, as $\phi_2(\pi)$ coincides with π in all cases. On the other hand, ϕ_3 is important. This function evidently describes the behaviour of the propositional compound[17] 'not *A*' in respect of the correlated truth-values. If we employ the symbol '$\neg$', first used for 'not *A*', also for ϕ_3, we may write the matrix in question as follows

π	$\neg(\pi)$
T	**F**
F	**T**

or briefly

	$\neg$
T	**F**
F	**T**

The use of the sign '$\neg$', which was introduced as abbreviation for 'not', as a sign for the corresponding truth-function is justified by the fact that '$\neg(\pi)$' and '$\neg A$' are not likely to be confused and also by the consideration that the abbreviated form of the matrix may be regarded as a direct description of the 'truth-behaviour' of the propositional compound '$\neg A$'.

Of the 16 dyadic truth-functions those again are trivial whose value does not depend on both arguments (i.e. depends only on one or on neither). If these are excluded, we are left with the 10 following functions, which are here numbered purely for convenient reference in this chapter.

π	ρ	$\psi_1(\pi,\rho)$	$\psi_2(\pi,\rho)$	$\psi_3(\pi,\rho)$	$\psi_4(\pi,\rho)$	$\psi_5(\pi,\rho)$
T	**T**	**T**	**T**	**T**	**T**	**T**
T	**F**	**F**	**T**	**F**	**T**	**F**
F	**T**	**F**	**T**	**T**	**F**	**F**
F	**F**	**F**	**F**	**T**	**T**	**T**

π	ρ	$\psi_6(\pi,\rho)$	$\psi_7(\pi,\rho)$	$\psi_8(\pi,\rho)$	$\psi_9(\pi,\rho)$	$\psi_{10}(\pi,\rho)$
T	**T**	**F**	**F**	**F**	**F**	**F**
T	**F**	**T**	**F**	**T**	**F**	**T**
F	**T**	**T**	**F**	**F**	**T**	**T**
F	**F**	**T**	**T**	**F**	**F**	**F**

Like the function ϕ_3, the above functions characterize situations that occur under the most varied conditions of our environment and are therefore important also from the point of view of the idealized world referred to in II 1, p. 28. If we wish to describe them, we generally have to use the propositional connective that corresponds truth-functionally to the matrix in question. Most languages have their own words[18] for the most important of these connectives, and these words also serve to characterize the situations to which they are applicable. We shall here mention only a few examples relevant to English.

(ψ_1) Plants thrive if light *and* water are available in the correct quantities.[19] An electric lamp whose supply of current is regulated by two switches placed 'in series', will light if and only if switch 1 *and* switch 2 have the correct position. (This situation occurs e.g. if switch 1 is the 'master switch'.) Clearly, the common element of the two situations is described by ψ_1, and ψ_1 also describes the truth-functional behaviour of the propositional connective in '*A and B*'. We shall therefore use the same sign, viz '$\wedge$', both for this propositional connective and for its corresponding function ψ_1, and thus write "$A \wedge B$' and '$\wedge(\pi,\rho)$'.

(ψ_2) A two-link chain will break if the first *or* the second link breaks. An electric bell controlled by two 'parallel' bell-pushes will ring if the first *or* the second is pressed. We shall employ the sign '$\vee$' for the propositional connective in '*A or B*' (where 'or' is used as in these examples) and for the corresponding function ψ_2.

(ψ_{10}) *Either* we shall go to the theatre this evening *or* we shall watch the sun rise to-morrow (but we shall not do both). *Either* the child will have a building set for Christmas *or* an electric train (but not both). These situations are obviously described by ψ_{10}, and not by ψ_2.

The above ten functions can to some extent be systematized, as the following considerations will show. It is easily verified that on the basis of our numbering ψ_i of the functions, the following holds for $i=1, \ldots, 5$ and any values **T**, **F** for π, ρ:

$$\psi_{i+5}(\pi, \rho) = \neg(\psi_i(\pi, \rho)),$$

i.e. the last five functions can be obtained from the first five by interchanging **T** and **F** in all the spaces. Further, it is easily verified that the following holds in all cases:

$$\psi_4(\pi, \rho) = \psi_3(\rho, \pi) \quad \text{and} \quad \psi_9(\pi, \rho) = \psi_8(\rho, \pi)$$

The signs allocated below to the ten functions take account of these relationships.[20] With each *sign* we once more give the value distribution of the corresponding function, though this time in a different form, and also the characteristic English words that would occur in a corresponding compound proposition.[21]

$\wedge$	T	F
T	T	F
F	F	F

...and...

$\vee$	T	F
T	T	T
F	T	F

...or... (or both)

$\rightarrow$	T	F
T	T	F
F	T	T

if... then... or ...only if...

$\leftarrow$	T	F
T	T	T
F	F	T

..., if...

$\leftrightarrow$	T	F
T	T	F
F	F	T

...if and only if...

$\overline{\wedge}$	T	F
T	F	T
F	T	T

it is not the case that both ...and... (by analogy termed the *nand*-function)

$\overline{\vee}$	T	F
T	F	F
F	F	T

neither... nor... (sometimes termed the *nor*-function)

$\overline{\rightarrow}$	T	F
T	F	T
F	F	F

..., but not...

$\overline{\leftarrow}$	T	F
T	F	F
F	T	F

not ..., but...

$\overline{\leftrightarrow}$	T	F
T	F	T
F	T	F

either... or... (but not both)

An analogous treatment of truth-functions with more than two arguments is obviously out of the question. We shall here mention only the fact that every truth-function can be built up using merely monadic and dyadic functions. The problem of finding the simplest – and hence the most economical – way of doing this, has in recent years become of great practical significance, since the processes carried out by modern computers can to a large extent be described in terms of truth-functions.

Some n-adic truth-functions are, however, important from the point of view of representing relations between concepts; thus for every n:

(1) The function that assigns the value **T** to that n-tuple consisting solely of **T**'s and the value **F** to all other n-tuples.

This function evidently describes the truth conditions of a propositional compound that asserts the simultaneous truth of *all* the members of a series of propositions A_i. Since for $n=2$ we obtain $A_1 \wedge A_2$, a propositional compound of this kind is often designated by '$\bigwedge_{i=1}^{n} A_i$'. The function[22] is therefore designated by '$\bigwedge_{i=1}^{n} \pi_i$'.

(2) The function that assigns the value **T** to n-tuples containing *at least one* **T**, and the value **F** to that n-tuple consisting exclusively of **F**'s.

This function evidently specifies the truth-conditions of a propositional compound which asserts the truth of at least one member of a series of propositions A_i. Since for $n=2$ we obtain $A_1 \vee A_2$, a propositional compound of this type is frequently designated by '$\bigvee_{i=1}^{n} A_i$'. As above we designate the corresponding function by '$\bigvee_{i=1}^{n} \pi_i$'.

Taking the case where the universe of discourse introduced in II 1, p. 29, is finite, the objects a_i, of this universe can be arranged in a finite series $(a_1, \ldots, a_N)$. By making use of the functions introduced above, we can now express relations between attributes, that do not refer to specific objects e.g.

$$\bigwedge_{i=1}^{N} (A^1 a_i \rightarrow B^1 a_i).$$

From a *formal* point of view this proposition asserts that the value **T** occurs in every place of the N-tuple$\rightarrow(A^1(a_i), B^1(a_i))$. This is so automatically for those i for which $\neg A^1 a_i$,[23] since $\rightarrow(\mathbf{F}, B^1(a_i))=\mathbf{T}$ according to the table for $\rightarrow$. However, for those i for which $A^1 a_i$ the condition $\rightarrow(\mathbf{T}, B^1(a_i))=\mathbf{T}$ must be fulfilled. According to the table for $\rightarrow$ this is possible only if B^1 (a_i) is **T**, i.e. only if $B^1 a_i$. It therefore holds for all a_i: if $A^1 a_i$ then (also) $B^1 a_i$, or briefly: all A^1-things are B^1-things. This is the *content* of the proposition we have been considering. It thus represents a new version of propositions of the type $A^1 \mathbf{a} B^1$ (as in I 2, p. 14) and at the same time interprets them, for 'finite worlds', in terms of a discrete ontology.

It is somewhat easier to see that $\bigvee_{i=1}^{N} (A^1 a_i \wedge B^1 a_i)$ asserts that at least one A^1-thing is a B^1-thing. This gives us an interpretation in terms of a discrete ontology of propositions of the type $A^1\mathbf{i}B^1$ (as in I 2, p. 14).

However, the scope of these interpretations is necessarily limited by the fact that they hold only for universes consisting of a fixed number of objects. The abbreviations '$\bigwedge_{i=1}^{N}$ ...' and '$\bigvee_{i=1}^{N}$ ...' stand for formulae that increase in length as N becomes larger, and in the case of a universe consisting of an infinite number of objects, they would have to be taken as representing 'infinitely long' formulae. Such a theory, as presented by C. Karp [1] is anything but elementary. For this reason two logical functions of a different kind are introduced, which assign truth-values to the monadic attributes themselves[24] on the basis of the following stipulations:

$\bigwedge(A^1) = \mathbf{T}$, if A^1 holds for all objects in the domain in question,
$\bigwedge(A^1) = \mathbf{F}$, in all other cases.
$\bigvee(A^1) = \mathbf{T}$, if A^1 holds for at least one object in the domain in question,
$\bigvee(A^1) = \mathbf{F}$ otherwise.

In order to be able to apply these functions, also to attributes defined by compound conditions, we write e.g.:

'$\bigwedge x(A^1 x \to B^1 x)$' or '$\bigvee x(A^1 x \wedge B^1 x)$'.

This may be interpreted as follows: $\bigwedge$ holds for the attribute $x(A^1 x \to B^1 x)$,[25] whose value 'for x' is given by $\to(A^1(x), B^1(x))$; and $\bigvee$ holds for the attribute $x(A^1 x \wedge B^1 x)$, whose value 'for x' is given by $\wedge(A^1(x), B^1(x))$.

It is easy to see that within a finite world consisting of N things $a_1, \ldots, a_N$

$$\bigwedge x(A^1 x \to B^1 x) \quad \text{coincides with} \quad \bigwedge_{i=1}^{N} (A^1 a_i \to B^1 a_i),$$

and

$$\bigvee x(A^1 x \wedge B^1 x) \quad \text{with} \quad \bigvee_{i=1}^{N} (A^1 a_i \wedge B^1 a_i).$$

We thus read:

'$\bigwedge x$ ...' as 'for all x ...',
'$\bigvee x$ ...' as for at least one x ...'.

Although the term 'quantifier' is in use both for these operators and the corresponding functions (see above) we shall adopt Hilbert's corresponding term '*quantor*' solely for the functional version, i.e. $\wedge$, $\vee$.

3. SOME CRITICISMS OF IDEALIZATION. CLOSED AND OPEN CONCEPTS

Various objections have been raised against the ontological foundation of logic outlined in the preceding chapter. These involve among other things, the question why expressions like 'not', 'and', 'or', 'all', 'at least one' should count as characteristic of the *form* of propositions, whereas expressions like 'apples', 'pears', 'potatoes', 'is sweet', 'is yellow' are regarded as variable. We shall refer to expressions of the first group as 'form expressions', distinguishing them from those belonging to the second group, which we shall call 'content expressions'.

It seems to us that the structures described by form expressions are so clearly delimited by virtue of their generality that they warrant the special investigation of their underlying regularities, and this is precisely what a logical investigation does. Further, the general validity of schemata such as the one discussed in II 1, (A) p. 28 depends on specific form expressions, as will be seen from the following examples, where one form expression has been altered in each case (the altered expression being shown in italics):

(B) *No* *A*-thing is a *B*-thing or there is at least one *A*-thing that is not a *B*-thing.

This schema is invalid, as e.g. the following is false: 'No murderer is a criminal or there is a murderer who is no criminal.'

(C) Every *A*-thing is a *B*-thing *and* there is an *A*-thing that is not a *B*-thing.

Every proposition subsumed by schema (C) is false.

(D) Every *A*-thing is a *B*-thing or *one knows* an *A*-thing that is not a *B*-thing.

One would surely be mistaken to infer that all electors will vote for party *X* from the fact that one does not know an elector who has not voted for party *X*.[26]

(E) Every *A*-thing is a *B*-thing or there is an *A*-thing that is *also* a *B*-thing.

The following, for example, is false at the present time: 'Every Englishman is an inhabitant of Mars or there is an Englishman who is also an inhabitant of Mars.'

A further objection – once it is admitted that idealization is a necessary prerequisite for the theoretical knowledge of a subject-matter – concerns the question whether the idealization described in II 1, p. 28 is adequate, i.e. whether it suppresses features of reality that are essential from the point of view of logic. After all, material existence itself is an abstraction: an apple, for example, usually loses its material existence very quickly and a china plate, say, is liable to lose it. This time factor is absent from our idealized world as it stands. It can, however, be introduced e.g. by regarding all things as things existing at a specific time. Between two things in the new sense, which 'are' one and the same thing in the old sense there then exists a peculiar relationship, designated as genidentity (after Kurt Lewin). In more abstract cases such genidentity is often 'created' by the introduction of a name, as when e.g. we refer to the 'eight-o'clock train' irrespective of whether it is made up of the same carriages each day and regardless even of a change in timetable according to which it leaves at 7.58 instead of at 8.07 as previously.

As against the relative endurance of phenomenal complexes, from which the notion of the existence of things has been abstracted, we have the relatively unlimited divisibility without corresponding loss of characteristic features of certain things in our environment. For example, we usually talk about liquids as though any partial quantity were equivalent in kind to the whole. In many cases the corresponding nouns have no plural form or have one only in a transferred sense as in:

'all oils'	=	'all types of oil',
'many coals'	=	'many pieces of coal',
'some whiskies'	=	'some kinds of whisky'
		but also
	=	'some measures of whisky'.

The convention whereby, in a variety of circumstances, we refer only to quantities that are whole multiples of a basic quantity, is in practice a very useful method of dealing with 'continuously divisible' things within the framework of a discrete ontology. 'I'll have a sherry'[27] is, for example, more intelligible than 'I'll have the locally traditional quantity of sherry in a locally traditional kind of drinking-vessel'. Continuously divisible quantities can be given a basis of strict proof in terms of a discrete ontology, but this involves considerable logical complications (in particular the introduction of so-called measure numbers[28]). A genuinely continuous ontology, which would form a counterpart to a discrete ontology, has, so far as we are aware, not yet been developed as a basis for logic. The beginnings of such an ontology can perhaps be seen in some of the attempted interpretations of wave mechanics in modern physics, which explain the occurrence of discrete phenomena in a world presupposed as continuous. However, these explanations do not amount to a truly continuous ontology, since the setting up of a 'wave equation' involves special physical assumptions, and also requires a mathematical conceptual apparatus which is based – ultimately – on a discrete ontology.

Finally we shall deal with objections connected with the fact that the compilation of a catalogue $\mathbf{C}_2$ presupposes a fixed domain of objects. An idealization of this kind is appropriate for cases where a finite domain of objects can be stipulated by convention. But difficulties arise, for example, when properties that are meaningful in respect of a specific domain of things are transferred to a larger one.

Let us suppose 100 apples, e.g. in a basket. They will have 2^{100} properties,[29] i.e. 2^{100} catalogues $\mathbf{C}_{A^1}$ in the sense of II 2, p. 32. Every partial quantity that can be taken out of the basket involves the property of belonging to it: from the point of view of logic it suffices that a quantity *could* be taken out of the basket, since if a logical proposition is to be asserted for any property whatsoever, it must not be falsifiable even by the oddest examples. Two properties that determine the same quantity do not need to be distinguished logically – not, at any rate, so long as properties are not themselves regarded as things (cf. V 3, p. 100). However, out of these 2^{100} properties only a very few have sufficient practical value to be designated by a special name, e.g. those properties that indicate membership of specific kinds or other qualities such as sweet, sour, aromatic, ripe, worm-eaten etc. And such properties are further distinguished from the totality of 'anonymous' properties in a way that is not covered by the introduction of the attribute concept. Let us introduce an additional apple, no. 101. Each of the 2^{100} properties yields two new properties in the enlarged domain, viz one that holds for no. 101 and one that does not. On the other hand apple no. 101 will in general have to be judged in respect of the named properties in a perfectly definite manner. Any possible uncertainty that might arise in this connexion is at any rate quite different from the complete arbitrariness of the 'anonymous' properties. If a property is, say, verbally defined with reference to the totality of the domain to be extended, then extending that domain may also change that property as applied to former values. Such arbitrariness or anonymity is no doubt also the source of the difficulty of expressing the similarity of repeatable events in terms of logic.

The situation is similar in the case of relations. Let us suppose 24 plates in addition to the apples. Any correlation whatsoever between apples and plates (e.g. which apple is to be placed on which plate) involves a relation and hence an attribute. Taking into account the possibility that some apples may not be placed on any plate and some on several plates, and including all border-line cases, our example yields 100·24 independent decisions and thus 2^{2400} (a number consisting of 723 figures) possible relations[30], most of which are 'anonymous'. Again there are attributes which, e.g. in the case of the number of apples being increased, extend themselves 'naturally': thus, for example, if only sweet apples are to be placed on specific plates.

We shall designate such properties or relations as open, and shall refer to them as open attributes, thus generalizing the attribute concept. These are in general given by a linguistically formulated condition,[31] which determines an attribute in the previous narrower sense in every suitable domain of objects.

It is, of course, possible that among attributes in the new, extended sense there are such that do not make use of the new freedom. Thus the property of having been a sweet apple in a specific basket on 1st September 1967 is not altered in any way by the fact that other apples in the basket have since ripened or that further apples have been added. Such an attribute in the extended sense will be designated as closed. A closed attribute is yielded by every attribute A^1 in the restricted sense *via* the property of belonging to the domain for which A^1 was originally meaningful, and of having in addition the property described by A^1. We proceed similarly for relations between n things belonging to a domain for which an attribute A^n is given.

The distinction between open and closed attributes is of particular importance for mathematics. The various standpoints which are nowadays held regarding the foundations of mathematics may to a large extent be grouped according to whether attributes applicable to infinitely many objects (e.g. numbers) are regarded as closed, and if so, which ones. Many propositions about the members of the number series 0, 1, 2, ... can be understood and proved without the series being regarded as closed, e.g.:

For every given prime number[32] there is a greater prime number: the next one can be reached in a previously limited number of steps.

On the other hand, there are problems that can be meaningfully formulated only if the entire number series is regarded as 'available'. For an example, cf. VII 3 C, p. 134.

The relative determinateness of the terms *S*, *P* (and *M*) in traditional syllogistic suggests that these be restricted to closed attributes over the open domain of all things – open, at any rate, if 'thing' is here understood in its widest sense as 'object of consideration'. The customary restriction to non-complementary subject-terms thus expresses the cautious attitude only to argue within closed concepts. This means, however, that **n***P* can no longer be regarded as equivalent in status to *P*, and the reduction outlined in I 3, p. 25 f. no longer holds. However, it could be replaced by e.g.

$$(S - P) =_{\mathrm{Df}} [x \mid Sx \wedge \neg Px] \qquad \text{(cf. V 2, p. 96)}$$
$$S\mathbf{e}P =_{\mathrm{Df}} S\mathbf{a}(S - P)$$

and the corresponding auxiliary inferences.

NOTES

1 Why 'is sweet' is here treated as one word will be made clear in III 2, p. 55. Cf. also p. 31.

2 How to obtain an exact definition of this concept will be shown in III 3, pp. 57, 59, 61.

3 Note that this is a version of what is known as 'logical atomism'.

4 Strictly speaking a law of nature is more than a totality of experiences. These can only show a law to be very probable – in a peculiar sense of probability (cf. in this connexion VIII 4, p. 164 f.).

5 This characterization of mathematics is admittedly one-sided. Some other aspects of mathematics will be discussed in VII 2, p. 127 f.

6 The word 'Begriffsschrift' (which 'conceptual notation' is intended to translate) was first used by G. Frege in his conceptual system published in 1879. This expression appears to us to render our meaning better than the term 'formalized language' which nowadays has greater currency, but tends to evoke the connotation of ill-usage: there is no question, of course, of the language being spoken.

[7] In general the use of quotation marks in the following is intended to indicate that reference is made to the sign enclosed by them, cf. VI 2, p. 115.
[8] Here contrary to the convention laid down in note 7, reference is not made to the sign A_1 in conjunction with the sign n, but to the signs obtained from '$A_1{}^n$' (etc.) by substituting an appropriate numeral for 'n'. For a correct treatment of such *quasi-quotations* cf. Quine [2], p. 35.
[9] Cf. in this connexion also III 2, p. 55.
[10] More precisely, though hardly more intelligibly: the relation that holds between any persons or things x and y whatsoever in precisely those cases where x is taller than y.
[11] A superhuman power capable of compiling catalogue C_1 might also be credited with the compilation of the improved catalogne C_2; but in fact this latter task involves an essentially new element, cf. VII 2. However, it suffices to regard these catalogues as thought experiments, on which certain idealizations are to be based.
[12] The question whether the catalogue is to be regarded as consisting of expressions in the sense of linguistic structures or of their content, may be left open at this stage. On this distinction cf. also III 1, p. 63, note 3.
[13] A function f is given if every thing x out of a set S is ascribed exactly one thing 'by f'; this is designated as '$f(x)$'. S stands for the set of arguments or the domain of f.
[14] We have already had to speak about language on several occasions, e.g. when we have introduced a new linguistic expression by a definition and not through use. Certain problems of 'talking about language' will be discussed in III 3, p. 56 f. and in VI 2, p. 115.
[15] As the dyadic truth-functions are the most important practically, the n-adic ones are rarely dealt with individually.
[16] This rapid numerical increase of the n-adic functions is no doubt the main reason why interest in individual functions is slight.
[17] Although this is not a case of several propositions being combined, it is nevertheless convenient to subsume it under the general heading of compound propositions.
[18] Cf. in this connexion the paper by Döhmann [1].
[19] The reader might care to re-formulate this sentence so that 'and' is really used to connect two *propositions*.
[20] Such 'relationships' are best discussed in a linguistic formulation and will therefore be left for III 3, p. 58 f.
[21] The extent to which the truth conditions of these propositional connectives are expressed by their corresponding functions, and whether they can be rendered at all by a truth-function, will be discussed in IV 3, p. 78 f. We here regard the value distribution as primary, and the existence of an adequate linguistic formulation as a convenient extra.
[22] More precisely: its value for arbitrary n-tuples $(\pi_1, \ldots, \pi_n)$ of truth-values.
[23] To be read: A^1 does not hold for a_i.
[24] Instead of to the series of truth-values obtained by applying an attribute to each member of the series of things in turn.
[25] The special significance of the initial x is frequently expressed symbolically, e.g. by writing $\hat{x}$ ... *(Principia Mathematica)*, λx ... (Church [2]), $[x]$ ... (Cogan [1], p. 202). On the difference between $[x \mid \ldots]$ and $[x] \ldots$ cf. V 2, p. 94. On the need to distinguish between a function and a general functional value, cf. *ibid.*
[26] False inferences of this kind are, however, often made. This is perhaps explained by the fact that a refined form of this schema is admitted in the logic of probability (cf. VIII 3, R 3, p. 158). The above example, however, would not warrant a probability

inference, since one might consciously have restricted one's circle of acquaintances to people in favour of party *X*.

[27] The reader is asked to translate this himself into a locally traditional formula.

[28] Non-negative real numbers, i.e. those numbers that can be represented by (possibly infinite) decimal numbers (e.g. 1.35; 3.333 ...; 3.1415 ...) can be shown to be measure-numbers in terms of an expanded logic as in V 3.

[29] i.e. more than 10^{30}.

[30] i.e. catalogues C_{A^2} in the sense of II 2, p. 33.

[31] We refer here to something non-linguistic. However, if one tries to give examples, the linguistic formulation of such conditions becomes unavoidable.

[32] i.e. a positive integer divisible only by itself and by 1. For a proof, cf. for example Pólya [3], p. 192 f.

CHAPTER III

LOGIC AS LINGUISTIC THEORY

When we, consciously or unconsciously, apply a law of logic, this is simply a kind of *activity*. But if we wish to formulate or to prove a law of logic, then we cannot avoid *talking about* language, for the forms of propositions are vitally important for the formulation of logical laws. Even if we are primarily concerned with the formal properties of the *contents* of sentences, these are in general best discussed in linguistic terms (cf. the examples given in II 2, p. 36 to illustrate the structures described by 'and', 'or', 'either – or'). This gives rise to a linguistic theory which differs from a philological investigation of language mainly in that it abstracts to a large extent from the contingencies of linguistic development. The result is a standardization of language, and this standardization is then incorporated in a symbolic system, i.e. a symbolic notation.

1. THE FORMS OF PROPOSITIONS. GRAMMATICAL AND LOGICAL SYNTAX

In formulating logical laws we are essentially dependent on the form of propositions, as shown by the examples in I 1, p. 10. But when we wish to apply a logical law in order to make inferences from premises formulated in natural language, we very soon realize that we cannot always rely upon the grammatical form of propositions. This may be illustrated by the following example:

No cat has two tails.
One cat has one tail more than no cat.
―――――――――――――――――――――
One cat has three tails.

We shall not concern ourselves with the fairly crude misuse of language involved here. The ambiguity expressed in this example, however, is not unique, as the following examples will show.

(1) The train was 20 minutes late.

(2) The whale is a mammal.

(3) The meek shall inherit the earth.
(4) The truest friends are friends in adversity.
(5) The best plans go astray.
(6) The end does not justify the means.

Of the above examples (1) is a singular proposition, but (2), (3), (4) ('all who are friends in adversity are true friends') and (6) ('no cases are cases where the end justifies the means') are universal. The meaning of (5) seems to be 'Some good plans go astray'.

On the other hand one and the same logical form may be given such different linguistic expressions that in some cases it can be determined only from the context, as the following examples will show.

(7) The last mile is the most difficult.
(8) A miss is as good as a mile.
(9) Every dog has his day.
(10) Elephants never forget.
(11) Any suggestions will be welcome.
(12) He who laughs last laughs best.
(13) Anything will make a story.
(14) All the lights went out.

Each of the sentences (7) to (13) *can* be used in the sense of 'all ...', but e.g. (9) also in that of 'this ...' and (10) also in that of 'in general ...'.

Instructive as it is to look for the linguistic variants of logical forms, and useful as such variants are from the point of view of finding an acceptable formulation, it is nevertheless an essential requirement of logic that among the various equivalent (or near-equivalent) formulations of any one proposition, one should be selected as the standard form; thus e.g. in respect of examples (7) to (12),

'All *A*-things are *B*-things.'

In the same way standard forms may be assigned to other logically relevant linguistic formulations. Thus

'If *A*, then *B*'

may be regarded as standard form for '*A* only if *B*', '*A* implies *B*'[1] etc. However, even if the formulations chosen as standard forms are kept as concise as possible, the standardized forms of a complicated sentence will in most cases look clumsy and be practically unintelligible. For this reason special symbols[2] have been introduced for the standard forms of those natural-language expressions that are logically relevant. Once the relationship between the latter and the former has been discussed and agreed upon, logical laws can be formulated and investigated in terms of propositional forms, which can be expressed in a symbolic notation constructed in the main out of the symbols introduced for this purpose. However, it is the standardization of language that is of basic importance, although the introduction of a symbolic notation is at least equally significant from a practical point of view. The following example should make this clear:

Premise: Whenever *B* follows from *A*, then *C* (holds).

Assertion: Whenever *A* follows from *C*, then *A* (holds).

In this example the propositional connective to which we have above assigned the standard form 'if – then –', is expressed in three different ways, viz: (1) 'Whenever –, then –', (2) 'from – follows –', (3) *premise:* –, *assertion:* –'.

The standardized form would thus read:

If: : if: if *A*, then *B;* then *C*,
then: : if: if *C*, then *A;* then *A*.

This form would indeed tax the reader's patience. On the other hand, using the symbolic notation described in detail in III 2, p. 51 f. (where '$A \to B$' stands for 'If *A*, then *B*'), we obtain a well-formed and readily intelligible formula:

$$((A \to B) \to C) \to ((C \to A) \to A).$$

Such formulas are also known as *propositional forms*. It seems reasonable enough to say that the form of a proposition is that which two propositions of the same form have in common. This is not a circular definition, since despite the linguistic formulation the concept of 'having the same form' is more basic than that of 'having the form ...'. Set theory provides a possible definition: starting from the concept of 'having the same form', we define as the form of a proposition *P* the totality of propositions that have the same form as *P*.

In general, a totality thus defined is (or rather: determines) an open attribute in the sense of II 3, p. 42, since new propositions of the same form can usually be constructed if the language is expanded. Open attributes, however, cannot be treated in every sense as objects, as will be shown in VI 1, p. 114.

The following is a more elementary definition of the concept of form. (It stands in much the same relation to the set-theory concept of form as does a jelly mould to the concept of form defined as a spatial property belonging to all jellies moulded or yet to be moulded by it, or as the totality of these jellies.) We thus define as follows: the form of a proposition *P* is a formula constructed exclusively out of variables and logical expressions (or symbols standing for them), from which proposition *P* may be obtained by appropriate substitutions of names for object variables, of names of concepts (i.e. in general, predicates) for concept or attribute variables and of sentences for propositional variables.[3]

Taking this definition as it stands, every proposition has the form *p*, if *p* is a propositional variable. This interpretation could be avoided by amending the definition, but this is not necessary so long as 'the form of a proposition' is used only in expressions such as

All propositions of a specific form P are true.

This 'ontological' statement may be reformulated as a logical statement:

The propositional form P is generally valid.

Thus, for example, $((p\rightarrow q)\rightarrow r)\rightarrow((r\rightarrow p)\rightarrow p)$ is a generally valid propositional form, as will be shown in III 3, p. 57, with the aid of an exact definition of general validity. Further methods for obtaining generally valid propositional forms are given in IV 2, p. 73 f.

2. STANDARDIZATION AND SYMBOLIZATION

We are able to formulate in words complicated logical relationships by making use of the variety of linguistic expression, but this variety is a hindrance from the point of view of recognizing logical laws and can only be overcome by some sort of standardization. And since a language that has been merely standardized is unintelligible for practical purposes, we shall combine standardization with the introduction of a symbolic notation. For this purpose we shall use among others the symbols already used naively in chapters I and II.

We begin with some simple examples.

A. The language $\mathbf{L}_P$ of propositional logic is constructed out of certain *basic signs*.[4] We require:

the series of propositional variables $p_1, p_2, p_3, \ldots$;
the propositional-logic functors $\neg, \wedge, \vee, \rightarrow, \leftrightarrow$,[5]
the parentheses (,) as auxiliary signs.

Among the finite sequences (strings of signs, indicated by $Z, Z_1, Z_2, \ldots$) that can be formed from the basic signs, propositional-logic *formulas* (in order to avoid the connotation of validity: a *w*ell *f*ormed *f*ormula, or a *wff*) are characterized as follows:

(1) Every variable is a *wff*,
(2) if Z is a *wff*, then $\neg Z$ is also a *wff*,
(3) if Z_1, Z_2 are *wffs*, then $(Z_1 \wedge Z_2)$, $(Z_1 \vee Z_2)$, $(Z_1 \rightarrow Z_2)$, $(Z_1 \leftrightarrow Z_2)$ are also *wffs;*
(4) only what can be shown to be a *wff* under (1) to (3), is to count as a propositional-logic *wff*, or: a P-*wff*.

Thus e.g. the following are *wffs:* under (1), p_1, p_3, p_4; under (2), $\neg p_4$; under (3), $(p_3 \vee \neg p_4)$, $(p_1 \wedge p_3)$; and by use of these also under (3), $((p_1 \wedge p_3) \rightarrow (p_3 \vee \neg p_4))$.[6]

This still leaves undefined the relationship between *wffs* and the truth-functions introduced in II 2; this will be done in III 3.

B. A language $\mathbf{L}_\Sigma$ of syllogistic may be similarly described. Referring to the syllogistic outlined in I 3, p. 18 f.[7], the following *basic signs* are required:

the concept variables $P_1, P_2, P_3, \ldots$,[8]
the (concept-forming) functor **n**,
the (proposition-forming) functors **a**, **i**,
the propositional logic functors $\wedge, \rightarrow$,
the parentheses (,).

From among the sign strings that can be formed from these basic signs, concept *terms* are distinguished as follows:

(1) every variable is a term,
(2) if Z is a term, then **n**Z is also a term,
(3) only what has been formed in accordance with (1) and (2) is a term.[9]

Wffs of syllogistic may now be introduced, e.g. as follows:

(4) if Z_1, Z_2 are terms, then $Z_1\mathbf{a}Z_2$, $Z_1\mathbf{i}Z_2$ are *wffs*,

(5) if Z_1, Z_2 are *wffs*, then $(Z_1 \wedge Z_2)$ and $(Z_1 \rightarrow Z_2)$ are also *wffs*.

Wffs that do not contain *wffs* as parts are called *atomic wffs*. The reader should note that following the rules $\mathbf{n}P_1\mathbf{a}P_2$ is not built up from $P_1\mathbf{a}P_2$, which is part of the former, as a string. This inconsistency could be avoided by modified conventions requiring, say $(P_1\mathbf{a}P_2)$ and $(\mathbf{n}P_1\mathbf{a}P_2)$ or $\mathbf{a}P_1P_2$ and $\mathbf{an}P_1P_2$.

(6) Only what has been formed in accordance with (4) and (5) is a *wff* of syllogistic, or: a Σ-*wff*.

This allows the construction of *wffs* having the forms[10] $(Z_1 \rightarrow Z_2)$ and $((Z_1 \wedge Z_2) \rightarrow Z_3)$, which are required for the symbolic representation of the auxiliary propositions and propositions of syllogistic. It would, of course, be possible to restrict the formation of formulas to what is strictly necessary, thus:

(4′) If Z_1, Z_2 are terms, then $Z_1\mathbf{a}Z_2$, $Z_1\mathbf{i}Z_2$ are atomic formulas.

(5′) If Z_1, Z_2, Z_3 are atomic formulas, then the combinations $Z_1 \rightarrow Z_2$ and $Z_1 \wedge Z_2 \rightarrow Z_3$ are *wffs*.

(6′) Only what has been formed according to (5′) is a *wff*.

C. Probably the most important symbolic notation used in modern logic is the language $\mathbf{L}_F$ of functional or predicate logic.[11] This may be interpreted as the language of discrete ontology in the sense of II 1, (p. 28) and 2, and such an interpretation suggests the following specific structuring of the language. On the basis of the designations introduced in this connexion (cf. pp. 30, 35, 37, 39) the following *basic signs* are introduced:

object variables $a_1, a_2, a_3, \ldots,$

predicate variables[12] $A_1^1, A_2^1, A_3^1, \ldots, A_1^2, A_2^2, A_3^2, \ldots,$

in general: A_k^n, where n indicates the number of places and k is a distinguishing sign;

propositional variables $A_1^0, A_2^0, A_3^0, \ldots,$[13]

functional variables $f_1^1, f_2^1, f_3^1, \ldots, f_1^2, f_2^2, f_3^2, \ldots,$

in general: f_k^n, n again indicating the number of places and k being a distinguishing sign;

propositional logic functors $\neg, \wedge, \vee, \rightarrow, \leftrightarrow$;

predicate logic functors $\bigwedge, \bigvee$;

parentheses (,).

Using these basic signs we begin by forming (object) *terms*[14] – strings of signs to be used as names for objects:

(1) Every object variable is a term;

(2) if $Z_1, \ldots, Z_n$ are terms, then $f_k^n Z_1 \ldots Z_n$ is a term[15];

(3) only what has been formed according to (1) and (2) is a term.

n-place *predicates* are formed next[16]:

(4) every *n*-place predicate variable A_k^n is an *n*-place predicate;

(5) if Z_1 is an $(n+1)$-place predicate and if Z_2 is a term, then Z_1Z_2 is an *n*-place predicate. (Thus if *n* terms are added consecutively to an *n*-place predicate, a null-place predicate is formed.)

(6) Only what has been formed in accordance with (4) and (5) is a predicate.

Finally, predicate logic *wffs* are formed:

(7) every null-place predicate is a *wff* (i.e. an 'atomic formula'), thus e.g. A_3^0, $A_2^1 a_1$, $A_2^2 a_1 f_1^2 a_2 a_3$;

(8) if Z is a *wff*, then $\neg Z$ is also a *wff;*

(9) if Z_1, Z_2 are *wffs*, then $(Z_1 \wedge Z_2)$, $(Z_1 \vee Z_2)$, $(Z_1 \rightarrow Z_2)$ and $(Z_1 \leftrightarrow Z_2)$ are also *wffs;*

(10) if Z is a *wff*, then so are $\bigwedge a_i Z$ and $\bigvee a_i Z$[17];

(11) only what has been formed in accordance with (7) to (10) is a *wff*, or: a F-*wff*.

The general designation for terms, predicates and *wffs* is *expressions.* When talking about $\mathbf{L_F}$ terms will be indicated by s, t; predicates by P, Q; formulas by A, B, ...

One might say that clauses (1) to (11) determine the grammar of the language of predicate logic, since they state what is to count as a meaningful, i.e. interpretable expression. On the basis of the expressions formable under (1) to (10) grammatical categories can be introduced that correspond, in part at any rate, to natural language grammatical categories. Thus, monadic predicates correspond to intransitive verbs, dyadic predicates to verbs with one object, triadic predicates to verbs with two objects, etc. But there are also grammatical categories that are specific to the language of predicate logic. Thus it has become customary to classify and designate *F-wffs* as follows:

(A ∧ B) as conjunctions,

(A ∨ B) as disjunctions (or alternations or adjunctions),

(A→B) as conditionals (or implications or subjunctions),
(A↔B) as biconditionals (or equivalences),
$\bigwedge x$A as universal formulas,
$\bigvee x$A as existential formulas.

Wffs containing only monadic predicate variables constitute the language of *monadic* predicate logic. In general, this is understood as not including the use of monadic function variables, even though this extension would preserve most features of simplicity.

Conventions regarding the saving of parentheses in $\mathbf{L}_F$ may be usefully laid down as follows:

Outside parentheses may be left out.

Each subsequent sign in the sequence ∧, ∨, →, ↔ is weaker[18] than each preceding sign. A sign with dots (e.g.: .∧., :→:) is weaker than any sign with fewer dots. – But note that there are many different conventions about such 'preference rules' in the literature.

The language may be *interpreted* by translation into an already interpreted language, e.g. into natural language, variables being generally replaced by specific names or predicates (but cf. III 3, p. 56 f.). However, the variables in (10) represent a special case: because of the initial operator $\bigwedge a_i$ (or $\bigvee a_i$) all variables a_i in the subsequent scope A are bound by the operator (thus becoming 'bound variables') and cannot be interpreted. Another way of expressing this is to indicate the reference of the operator by writing:

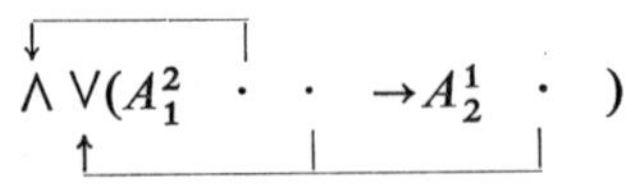

instead of: $\bigwedge a_i \bigvee a_2\ (A_1^2 a_1 a_2 \to A_2^1 a_2)$, thus leaving out the bound variables. In other words, the bound variables merely mark the places to which the preceding operator refers.

We can now deal with a formula where $\bigwedge a_1$ (or $\bigvee a_1$) occurs again in the scope of $\bigwedge a_1$ (or $\bigvee a_1$), as e.g. in

$$\bigwedge a_1(\bigvee a_1 A^2 a_1 a_1 \to A^1 a_1).$$

We transform in two steps:

$$\wedge a_1(\vee A^2 \;\cdot\;\cdot\; \rightarrow\; A^1 a_1),$$

$$\wedge(\vee A^2 \;\cdot\;\cdot\; \rightarrow\; A^1 \;\cdot\;).$$

It is evident that bound variables may be replaced by others without altering the sense of a formula, so long as the reference of the operator remains the same. Thus, for example, this reference would be altered in $\wedge a_1 A^2 a_1 a_2$ if we substituted a_2 throughout for a_1.

If object variables are replaced by names and functional variables by descriptive phrases,[19] then the terms – after any linguistic 'polishing' that may be necessary – link up into compound names, i.e. descriptions of objects.

The properties or relations indicated by predicate variables are generally expressed by verbs. For example, M^2ab might stand for *a* meets *b*, G^3abc for *a* gives *c* to *b*, L^3abc for *b* lies between *a* and *c*[20], E^4abcd for *a* exchanges with *b* article *c* for article *d*. And, to give a further example, $G^3f_1^1af_2^1ab$ gives the structure of: Harry's father gives Harry's wife a flower. The formation of compound predicates corresponds to the possibility of circumscribing properties and relations, for which there is often no verb in natural language.

Clause (5) is primarily a device for simplifying the formation of formulas, although it can also in some cases simplify that of predicates. For example, if G^3abc is interpreted as '*b* gives *a* to *c*' and if *a* stands for 'indemnity', then 'to indemnify' may be rendered by G^3a.

The usual translation for the propositional logic functor $\neg$ is 'not', but this often necessitates a change in word order. A change in word order can be avoided if we translate 'it is not the case that ...'.

The propositional logic functors are in general rendered by conjunctions or corresponding turns of phrase, in particular:

- ... $\wedge$ ... by: '... and ...', 'both ... and ...';
- ... $\vee$ ... by: 'or' (in a non-exclusive sense, as in the Latin 'vel', or in 'and/or');
- ... $\rightarrow$ ... by: 'if ..., then ...', also by: 'from the fact that ..., it follows that ...'[21], or by circumlocutions such as 'premise: ... assertion: ...';
- ... $\leftrightarrow$ ... by: '... if and only if ...', sometimes shortened to 'iff'.

$\mathbf{L}_F$ is often used with predicate *constants* having a fixed meaning, rather than with predicate variables. For example, if '**P***a*' stands for '*a* is a point', '**S***a*' for '*a* is a straight line', '**L***ab*' for '*a* lies on *b*' (**P** and **S** thus being monadic predicates and **L** a dyadic predicate), then the axiom of geometry that 'For any two points there is at least one straight line on which both points lie', can be represented symbolically by

$$\bigwedge a \bigwedge b(\mathbf{P}a \wedge \mathbf{P}b \rightarrow \bigvee c(\mathbf{S}c \wedge (\mathbf{L}ac \wedge \mathbf{L}bc))).$$

But since in modern axiomatics the space to which the axioms refer is in general not fixed, such 'constants' are rather a kind of restricted or specified variable.

3. RELATIONS BETWEEN LANGUAGE AND REALITY (SEMANTICS)

By translating formulas into a natural language (at first literally, then idiomatically), we rather hide the fact that the language of predicate logic is designed for an idealized world. This does not matter, provided that the meanings of the words occurring in the translation are precisely defined by special conventions. And this, after all, is a necessary condition for all meaningful and correct inference.

However, it is also possible to give a direct description, in the sense of an interpretation, of the relation between a symbolic notation, the 'object language', and an appropriate world. To do this we must, of course, be able to use the language in which this description is to be given, i.e. the 'metalanguage', to talk not only about the object language but also about the 'world' in question, so that there would seem to be little point in the whole procedure. In fact, however, such a description throws the relation of the 'object language' to the 'world' into greater focus – in much the same way as a silver spoon can be polished with a rag. See Heisenberg [1], p. 190.

A. The language $\mathbf{L}_P$ of propositional logic could be interpreted in terms of a 'world of (thinkable) states of affairs'. For every state of affairs s in such a world there would be a state of affairs s', which would consist in the fact that s does not obtain. And for every pair of states of affairs s, s' there would further be a state of affairs s'', consisting in the fact that *both* s *as well as* s' obtain, etc. As we have purposely restricted ourselves to compound propositions whose truth depends solely on the truth of their component parts[22], our interpretation may be made in terms of the truth values **T**, **F** – a procedure which will involve a greater degree of abstraction, but which will also be much simpler.

We thus interpret the functors (i.e. the signs) $\neg$, $\wedge$, $\vee$, $\rightarrow$, $\leftrightarrow$ in terms of the truth functions introduced in II 2, p. 35, 37 (and designated there by these signs.) Our procedure may be described as follows. Every assignment of truth values **T**, **F** to the propositional variables (in short: every assignment to the variables) also assigns a truth value to every formula A. This assignment $\mathfrak{B}^*$ determined by $\mathfrak{B}$ may be described step by step *via* the construction of A, viz as follows:[23]

$$\begin{array}{lll}
\mathfrak{B}^*(p_i) & = \mathfrak{B}(p_i) & \text{for all propositional variables } p_i \\
\mathfrak{B}^*(\neg A) & = \neg(\mathfrak{B}^*(A)) & \text{for the negation of a formula A,} \\
\mathfrak{B}^*(A \wedge B) & = \wedge(\mathfrak{B}^*(A), \mathfrak{B}^*(B)) & \\
\quad\vee & \quad\vee & \text{for combinations} \\
\quad\rightarrow & \quad\rightarrow & \text{of A and B.} \\
\quad\leftrightarrow & \quad\leftrightarrow &
\end{array}$$

Clearly, in order to determine $\mathfrak{B}^*(A)$, we need to take into account only the values assigned to the variables occurring in A. However, the *way in which* the value of a compound formula is determined by the values of the component parts, can be more simply described if we assume that $\mathfrak{B}$ makes assignments for all variables. If $\mathfrak{B}^*(A) = \mathbf{T}$ for all assignments, then A is *generally-valid* or a theorem of propositional logic, and we write symbolically: $\vDash_P A$.[24]

In order to determine whether a given P-*wff* A is a theorem, we proceed as follows. We note the value-assignments to the variables occurring in A, e.g. for

p	q	r	in $A = ((p \rightarrow q) \rightarrow r) \rightarrow ((r \rightarrow p) \rightarrow p)$
T	**T**	**T**	
T	**T**	**F**	
T	**F**	**T**	
T	**F**	**F**	
F	**T**	**T**	
F	**T**	**F**	
F	**F**	**T**	
F	**F**	**F**	
1	2	3	1 2 3 3 1 1
			4 5
			6 7
			8

The right-hand columns marked 1, 2 or 3 are filled in from the left-hand ones. The remaining columns are filled in according to the table for →: first 4 (from 1 and 2) and 5 (from 3 and 1), then 6 (from 4 and 3) and 7 (from 5 and 1), and finally 8 (from 6 and 7). If 8 is **T** in all cases, then A is a theorem. In the above example, **T** occurs at 8 for all assignments in A.

A calculus in the sense of IV 2, p. 71 f. may be obtained by appropriately rationalizing the evaluation of the formulas.

If *wffs* A, B have the same course-of-value, then A↔B is generally valid. Formulas of this type may be used in particular to express the definability of functors (II 2, p. 37). For every course-of-value (i.e. for every formula A) there is a 'standardized' formula B with the same course-of-value (that is, ⊨A↔B), e.g. in the form[25]

$$\ldots \vee (s_1 \wedge \ldots \wedge s_i) \vee \ldots,$$

where s_j ($j=1, \ldots, i$) stands in each case for p_j or for $\neg p_j$. (Every assignment to the relevant variables satisfies exactly one conjunction of this kind.) B is known as the (in this case: disjunctive) normal form (here: of A).

In the following concepts based on interpretations will, where relevant, be included under '$\mathbf{L}_P$'.

B. The language $\mathbf{L}_\Sigma$ of syllogistic may be interpreted in terms of any 'world' **D** of objects and their properties. In an interpretation of this kind we are concerned only with the extensions of properties, since our intended interpretation of the proposition-forming functors depends only on these. In accordance with their intended meaning we interpret as follows:

n, as forming the complementary concept $\nu(P_i)$ from P_i in **D**;

$P_i\mathbf{a}P_j$, as the logical function α, having the value **T** if and only if the extension of P_i is a part of the extension of P_j;

$P_i\mathbf{i}P_j$, as the logical function ι, having the value **T** if and only if the extensions of P_i and P_j have a common part;

An assignment $\mathfrak{V}$ of properties (or their extensions) to the variables P_i, also determines values $\mathfrak{V}^*(T)$ for the terms T. We thus obtain the truth values

$$\mathfrak{V}^*(T_1\mathbf{a}T_2) = \alpha(\mathfrak{V}^*(T_1), \mathfrak{V}^*(T_2)),$$

or

$$\mathfrak{V}^*(T_1\mathbf{i}T_2) = \iota(\mathfrak{V}^*(T_1), \mathfrak{V}^*(T_2))$$

for the atomic *wffs*. We then proceed as in the case of propositional logic, with the exception that instead of general-validity it is natural first to define validity in a fixed domain **D**, or more concisely: **D**-validity. It can be shown that it is only the number of objects in **D** that is relevant in this connexion.

If *general-validity* is defined as validity in every domain, it can be shown that a Σ-*wff* A is generally valid if and only if it is valid in a finite domain of 2^n objects, where n is the number of variables occurring in A.[26] Thus the *wffs* by means of which the laws of syllogistic may be represented, are generally valid if they hold for a domain of $8=2^3$ objects.

The definition of general-validity may also be adapted to the traditional convention, which excludes empty properties. We need only restrict assignment to variables to non-empty properties. We then proceed as above.

C. The language $\mathbf{L}_F$ of predicate logic is interpreted similarly as under B. Validity is first defined in a domain **D** of objects. However, because of the greater expressive range of $\mathbf{L}_F$ the notion of assignment $\mathfrak{B}$ must here be extended. Each variable must be assigned suitable objects, thus:

object variables a_i, objects from **D**;
predicate variables A_k^n, n-place attributes over **D**;
propositional variables A_k^o, truth values (as 'null-place attributes');
functional variables f_k^n, functions 'of n variables in **D** with values in **D**'.

An assignment of this kind is thus an *assignment over* **D**.

Every assignment $\mathfrak{B}$ then determines:
the values $\mathfrak{B}^*(\mathsf{t})$ of all terms t,
the values $\mathfrak{B}^*(\mathsf{P})$ of all predicates P,
the values $\mathfrak{B}^*(\mathsf{A})$ of all formulas A.

These values are defined step by step *via* the construction of t, P and A, i.e. for terms as *objects* by means of

$$\mathfrak{B}^*(a_i) = \mathfrak{B}(a_i),$$
$$\mathfrak{B}^*(f_k^n\mathsf{t}_1 \ldots \mathsf{t}_n) = \mathfrak{B}(f_k^n)(\mathfrak{B}^*(\mathsf{t}_1), \ldots, \mathfrak{B}^*(\mathsf{t}_n));$$ [27]

for predicates as *attributes* by:

$\mathfrak{B}^*(A_k^n) = \mathfrak{B}(A_k^n)$ for $n > 0$;

$\mathfrak{B}^*(\mathsf{Pt})$, where P is an $(n+1)$-place predicate $(n>0)$, is that n-place attribute that holds for precisely those n-tuples $(\mathfrak{x}_1, \ldots, \mathfrak{x}_n)$, for which $\mathfrak{B}^*(\mathsf{P})$ holds for $(\mathfrak{B}^*(\mathsf{t}), \mathfrak{x}_1, \ldots, \mathfrak{x}_n)$; [28]

$\mathfrak{B}^*(x\mathsf{A})$ is that one-place attribute[29] which, for an arbitrary $\mathfrak{x}$ from **D**, holds for $\mathfrak{x}$ if and only if $((\substack{x\\ \mathfrak{x}})\mathfrak{B})^*(\mathsf{A})=\mathbf{T}$, where $(\substack{x\\ \mathfrak{x}})\mathfrak{B}$ is that assignment $\mathfrak{B}_0$ for which $\mathfrak{B}_0(x)=\mathfrak{x}$ and $\mathfrak{B}_0(v)=\mathfrak{B}(v)$ for all other variables v;

for *wffs* as *truth values* by:

$\mathfrak{B}^*(\mathsf{Pt})$, where P is a *one*-place predicate, is $\mathfrak{B}^*(\mathsf{P})$ applied to $\mathfrak{B}^*(\mathsf{t})$, i.e. a truth value;[30]

$\mathfrak{B}^*(A^\circ_k)=\mathfrak{B}(A^\circ_k)$;

$\mathfrak{B}^*(\neg\mathsf{A})$, $\mathfrak{B}^*(\mathsf{A}\wedge\mathsf{B})$, $\mathfrak{B}^*(\mathsf{A}\vee\mathsf{B})$, $\mathfrak{B}^*(\mathsf{A}\rightarrow\mathsf{B})$, $\mathfrak{B}^*(\mathsf{A}\leftrightarrow\mathsf{B})$ are reduced to $\mathfrak{B}^*(\mathsf{A})$ and $\mathfrak{B}^*(\mathsf{B})$, exactly as in propositional logic;

$\mathfrak{B}^*(\bigwedge\mathsf{P})$ is **T**, if the one-place attribute $\mathfrak{B}^*(\mathsf{P})$ holds for all objects in **D**, and otherwise **F**;

$\mathfrak{B}^*(\bigvee\mathsf{P})$ is **T**, if the one-place attribute $\mathfrak{B}^*(\mathsf{P})$ holds for at least one object in **D**, and otherwise[31] **F**.

In this way a value is determined for every *wff* A in respect of an assign- $\mathfrak{B}$. We can say that $\mathfrak{B}$ *satisfies wff* A in the case where $\mathfrak{B}^*(\mathsf{A})=\mathbf{T}$. This may be understood as follows: A expresses requirements, to be satisfied by $\mathfrak{B}$, in respect of possible states of the 'world'. *Wffs* expressing requirements in respect of $\mathfrak{B}(x)$ are indicated by $\mathsf{A}(x)$ etc.

(Again) A is *valid* in respect of **D** if $\mathfrak{B}^*(\mathsf{A})=\mathbf{T}$ for all assignments over **D**, i.e. where all assignments over **D** satisfy A. And finally A is *generally-valid* or a theorem of predicate logic, if A is valid in respect of **D** for every non-empty[32] domain **D** – or equally, if A is satisfied[33] by all assignments (over any non-empty domains whatsoever); symbolically: $\models_{\mathbf{F}}\mathsf{A}$. (Again) '$\mathbf{L}_{\mathbf{F}}$' should be understood as also including concepts based on interpretations.

In the case of predicate logic, unlike that of propositional logic and of syllogistic, the definition of general validity yields no general method for determining *that* a formula A is a theorem, nor does it yield a method for deciding *whether* A is a theorem. Only in the case where the decision can be reduced to a finite number of steps, can it be made: e.g. it can always be determined in respect of a *finite* **D** *whether* a formula is valid for **D**. The reason for this is that there is only a finite number of value assignments over a finite domain **D** for the finite number of variables occurring in a formula A.

We shall discuss some methods for determining *that* A is a theorem in IV. The limits of such methods and – in a sense – of all thinkable methods will be discussed in VII.

We must now draw the reader's attention to an important generalization of the concept of general validity. In many cases one requires a kind of validity in respect of a specific set $\mathscr{S}$ of value assignments: let A be *valid* in respect of $\mathscr{S}$ in the case where $\mathfrak{B}^*(\mathsf{A}) = \mathbf{T}$ for all $\mathfrak{B}$ from $\mathscr{S}$.[34] We consider first the case where $\mathscr{S}$ itself is stipulated as the set of value assignments $\mathfrak{B}$, that satisfy a formula B (i.e.: $\mathfrak{B}^*(\mathsf{B})=\mathbf{T}$). Then A is valid in respect of $\mathscr{S}$ if every value assignment that satisfies B also satisfies A. We thus say that A *follows from* B, or in symbols: '$\mathsf{B} \models_{\mathrm{F}} \mathsf{A}$'.[35]

From the table for $\rightarrow$ it is evident that

(1) $\mathsf{B} \models_{\mathrm{F}} \mathsf{A}$ if and only if $\models_{\mathrm{F}} \mathsf{B} \rightarrow \mathsf{A}$.

For both sides of this equivalence are falsified if and only if there is an assignment $\mathfrak{B}$ where $\mathfrak{B}^*(\mathsf{B})=\mathbf{T}$, $\mathfrak{B}^*(\mathsf{A})=\mathbf{F}$. By introducing this special *relation of consequence* we do not arrive at any essentially new means of expression but merely gain a shift in emphasis: the formulation in terms of 'follows' lays greater stress on the value assignments that satisfy B. The fact that the relation of consequence is a kind of generalization of general validity may be explained as follows. For generally-valid A's, B follows from A if and only if B is generally-valid, or in symbols:

(2) If $\models_{\mathrm{F}} \mathsf{A}$, then: $\mathsf{A} \models_{\mathrm{F}} \mathsf{B}$ if and only if $\models_{\mathrm{F}} \mathsf{B}$.

These and similar theorems relating the concepts of general-validity and of consequence also argue in favour of using the same symbol.

Theorem (1) indicates a close connexion between 'if –, then –' and 'from – follows –'. But it also shows a difference: the former *connects* propositions or *wffs*, the latter *talks about wffs* (and, clearly, the schema can also be used to talk about propositions).

Let us now consider the more important case, where $\mathscr{S}$ has been specified as the set of assignments that satisfies simultaneously all formulas of a given *set* of formulas S. Where S is finite this case involves nothing essentially new, since S may be reduced to a single equivalent formula. (For example, if $\mathsf{S}=\{\mathsf{B}_1, \mathsf{B}_2, \mathsf{B}_3\}$, then $\mathsf{B}=((\mathsf{B}_1 \wedge \mathsf{B}_2) \wedge \mathsf{B}_3)$ yields the same $\mathscr{S}$). Independently of this restriction we now define

'*from* S *follows* A', in symbols '$\mathsf{S} \models_{\mathrm{F}} \mathsf{A}$'

by:

(3) Every assignment that satisfies all formulas in S, also satisfies A.

We give below, without proofs[36], the most important inferences from this definition:

(4) If A belongs to S, then A follows from S.

(5) If A follows from S, then A similarly follows from any set that includes S.

(6) If A follows from S, and if B follows from S and A together, then B follows from S alone.

Theorem (6) is a generalization of the non-trivial components of theorem (2), as will be readily apparent if it is formulated in symbols:

(6′) If $S \vDash_F A$ and $S, A \vDash_F B$, then $S \vDash_F B$.

A much deeper significance attaches to the following theorem, which expresses what is probably the most important feature of the predicate logic concept of consequence (cf. Gödel [1], whose theorem X contains the kernel of this theorem.)

(7) If A follows from S, then A follows from a suitable finite sub-set of S.

For similarly as here for predicate logic, the concept of consequence may be defined also for the extensions of predicate logic that will be discussed in V 3, p. 100 f., as well as for the simpler symbolic languages outlined above. And it is precisely as far as there exist valid analogues to (7) that it has been possible to give adequate descriptions of the respective concepts of consequence by formal *methods* of proof.[37]

Concepts that have to do with the relations between linguistic structures and their meanings – such as '$\mathfrak{B}$ *satisfies* A', '$\mathfrak{B}^*(A) = \mathbf{T}$' – or that are defined in terms of linguistic structures and with reference to meanings – such as 'A is *generally-valid*', '*from* S *follows* A' – are usually designated as semantic[38], even if the meanings are idealized as much as in the definitions of this paragraph.

Such idealization is, however, unavoidable if we are to ask meaningfully whether a proposition – or a *wff* at a specific value assignment –

is true or false. For example, in the sentence 'Paul's brother has measles', we understand the term 'Paul's brother' only when we gather from the context which Paul is meant, and only if this particular Paul has exactly one brother (the one to whom reference may here be made); and the predicate 'has measles' becomes meaningful only when it is completed by a time specification. Similarly a scientific proposition as customarily formulated becomes a proposition in the strict sense only when the missing details are supplied by tacit or explicit conventions or by the context. This includes the distinction of different meanings of the same word, such as the literal and figurative meaning, or changes in the meaning of a word due to the passage of time. In any case, in any *one* context each word must have only one meaning[39], so as not to allow 'inferences' such as[40]:

All cunning people are foxes.
All foxes have four legs.

All cunning people have four legs.

or 'definitions' such as:

'There is no number whose square is a negative number. Such numbers are called imaginary numbers, and we use them in calculations according to the following rules: ...'[41]

Here the meaning of the word 'number' has clearly changed a little too rapidly, for the intention is, in fact, to introduce a new numerical concept.

NOTES

1 Cf. in this connexion our remarks in III 3, p. 61.
2 Such as those already used in II 2, p. 37, 39.
3 Thus variables are *used* to refer to objects, concepts and states of affairs, but are *replaced* by the linguistic or symbolic description of those designata (by names predicates and sentences).
4 Some arbitrariness is unavoidable in this connexion, but the reader will quickly develop a sense for what is essential.
5 This selection from the possibilities indicated in II 2, p. 35 f., is arbitrary, but can be justified. It consists of the only non-trivial monadic functor and of the 'positive' functors among the non-trivial dyadic ones (i.e. Ψ (T, T) = T.) Of these $\leftarrow$ is superfluous, as it can be trivially expressed by means of $\rightarrow$.
6 On reducing the number of brackets by the use of 'preference rules', cf. p. 54.
7 The reader should interpret analogously all the variants occurring in I.
8 We thus avoid the restriction to *S*, *M* and *P*.
9 According to these requirements all (concept) terms have the form $\mathbf{n} \ldots \mathbf{n}P_i$. We nevertheless give the general formulation, as this continues to hold even when further methods of term formation are introduced, such as e.g. the formation of (Z_1Z_2) from the terms Z_1, Z_2, which is to be read in much the same way as example (7) on p. 96.

10 On reducing the number of parentheses, cf. p. 60.
11 'Functional' here refers to those logical functions we introduced under the name of attributes. 'Predicate' is often used for attributes as well as for their symbolic representations. Thus 'functional' usually does not indicate the occurrence of explicit symbols of object-to-object functions in the language.
12 By analogy with 'object variables' we should really talk about 'attribute variables', but this would be to deviate too far from what has become established practice.
13 It is often useful to have propositional variables available in predicate logic. This may be done as suggested here, by treating them as null-place predicate variables. In this way convenient formulations may be obtained, such as e.g. clause (5) on p. 53.
14 These are extremely useful. They correspond to natural-language expressions such as 'Harry's father', 'the sum of 2 and 3', 'Harry's journey from London to Edinburgh' (in the case where this is regarded as one object [of thought]). Often, however, (object) terms are introduced only at a later stage, by way of an expansion of predicate logic, as in V 1, B, p. 91 f.
15 The dots to indicate the intervening expressions could have been avoided here as in (4) to (6).
16 These are linguistic structures that express properties or relations, or in general: attributes. Here we are really concerned with predicate *forms*, but we shall use the shorter designation. It would also be possible to introduce as predicates the structures $a_i Z$ or $[a_i]Z$ or $[a_i|Z,]$ exemplified in II 2, p. 39, but the methods that would be involved are not generally regarded as belonging to predicate logic, cf. in this connexion V 2.
17 Thus in each case two separate operations are merged. According to II 2, p. 39, we would have had to form one-place predicates $a_i Z$ from Z, and then to ascribe to these the property expressed by $\bigwedge$, or respectively $\bigvee$. We shall return to this possibility in V 2, p. 94 f.
18 In the same sense as that in which, in algebra, $+$ is weaker than $\cdot$, so that $a + b \cdot c$ is read as $a + (b \cdot c)$.
19 As in note 14.
20 As this example shows, it is a matter of expedience whether the variable standing for the middle object in the arrangement is allocated the middle position.
21 In the similar but shorter phrase 'from ... follows ...', the compound sentence is made up, not of sentences, but of names of sentences. The same applies to the phrase '... implies ...'. In connexion with these phrases cf. also III 3, p. 61.
22 Cf. I 2, p. 34.
23 In the following equivalences the propositional logic functors occur on the left as components of formulas, whereas on the right they are used meaningfully as designators of the truth functions introduced in II 2, p. 35, 37. In this way the co-ordination to be established by this definition is in a sense presupposed, but on the other hand, the definition is more easily remembered in this form.
24 The symbol '$\models_P$' thus does not belong to the language $\mathbf{L}_P$ of propositional logic, but to the language in which we speak *about* $\mathbf{L}_P$.
25 Theorems such as $(p \wedge q) \wedge r \leftrightarrow p \wedge (q \wedge r)$ and $(p \vee q) \vee r \leftrightarrow p \vee (q \vee r)$ suggest the introduction of rules for the omission of parentheses more advanced than those in III 2, p. 54.
26 For a proof cf. Scholz-Hasenjaeger [1], p. 212.
27 Thus, for example, if $\mathbf{D}$ is a domain of numbers, $\mathfrak{B}(f_1^2)$ addition, $\mathfrak{B}(a_1) = 3$, and $\mathfrak{B}(a_2) = 5$, then $\mathfrak{B}^*(f_1^2\, a_1 a_2) = 3 + 5 = 8$.

[28] If e.g. $\mathfrak{V}(A_1{}^2)$ is the relation of being smaller than, and if $\mathfrak{V}(a_1) = 3$, then $\mathfrak{V}^*(A_1{}^2\, a_1)$ is the property of being greater than 3. For one-place predicates, cf. below.
[29] This definition is *in this context* merely preparatory for the definitions of $\mathfrak{V}^*(\bigwedge x\mathrm{A})$ and $\mathfrak{V}^*(\bigvee x\mathrm{A})$ by means of $\mathbf{P} = x\mathrm{A}$. But cf. the generalization in V 2, p. 95.
[30] This could also be regarded as the formation of a null-place attribute out of a one-place attribute and an object; but this would be somewhat artificial and probably no simplification.
[31] That is, if $\mathfrak{V}^*(\mathrm{P})$ holds for no object in **D**.
[32] This customary restriction to non-empty domains probably reflects the traditional exclusion of empty concepts. In fact, however, such restriction is superfluous; since there can be no assignments in the above sense over empty domains, every formula is valid for empty domains according to our definition, but is uninteresting. The situation is somewhat different if we introduce assignments restricted to the 'free variables' of a *wff*. See Hailperin [1], Schneider [1].
[33] Here the domain **D** is really required only for the definition of the concept of value assignment: all attributes and functions given by $\mathfrak{V}$ must 'operate' over the same domain, to which must also belong the objects given by $\mathfrak{V}$. Apart from this, however, **D** merges into the interpretation of $\bigwedge$ and $\bigvee$.
[34] Since to every value assignment $\mathfrak{V}$ there corresponds a specific domain **D**, our above definition of $\mathfrak{V}^*$ will hold here in a similar sense. Cf. note 33.
[35] The sequence of symbols 'B, A' is chosen here so as to agree with that in the theorem below.
[36] For the proofs, cf. for example, Scholz-Hasenjaeger [1] §§ 33, 105, 113.
[37] Cf. in this connexion IV 3, p. 81, 84.
[38] After the Greek σημαίνειν (semainein) = to mean, designate.
[39] Or else it must be made clear that despite sounding and being written in the same way, different words are 'really' involved. This is in general not contested when words sound alike but are written differently, but becomes doubtful when the spelling is the same, and in particular if subtle shades of meaning are involved that can be distinguished only from the context.
[40] After Aebi [1], p. xvi f., also p. 320, where it is given as an example for a more seriously false inference.
[41] The shift in meaning has been intentionally contrived here, after many similarly challengeable formulations. Cf. L. Euler, *Algebra*, part II, sect. 1, chap. 10, § 149.

CHAPTER IV

LOGIC AS METHODOLOGY

In discussing the figures of the syllogism in I, we introduced the reader to a system of methods for obtaining theorems from other (already established or assumed) theorems, which itself could be interpreted as a system of logical theorems. In III 3 we then put forward as 'theorems of predicate logic' a system of theorems about the idealized 'world' described in II, but without obtaining any methods for the production of theorems. We shall now concern ourselves with such methods. These will in general take the form of rules of inference. If we leave out of account for the moment the motivation of such rules, we may regard them as sets of instructions for the production of linguistic structures from given linguistic structures, so that they can be described in much the same way as the 'formalized' languages $\mathbf{L}_{\mathrm{P}}$, $\mathbf{L}_{\Sigma}$ and $\mathbf{L}_{\mathrm{F}}$ in III 2. In Chapter VII we shall ask what can be said about such 'thought processes' in the light of the fact that they may be regarded as sets of instructions whose applicability must in every case be verifiable in a finite number of steps.

1. THEOREM LOGIC AND RULE LOGIC

As auxiliary science for other, in particular the deductive sciences, logic should above all be a system of rules of inference. But it should be a system: not a mere accumulation of sets of instructions. And these rules must be given a foundation of proof. These requirements may be met by the introduction of 'higher-order' rules, by means of which all rules are reduced to specific, if possible 'especially intuitive' basic rules. The 'higher-order' rules, too, should be intuitive.[1] The introduction of higher-order rules can be avoided by reducing all original rules to theorems, perhaps with the aid of *one* suitable rule (cf. p. 67), and then systematizing these theorems by means of new basic rules, corresponding to the earlier higher-order rules. In the process, of course, the original basic rules become 'basic *theorems*' and we obtain a semantic foundation of logic (to the extent that it is 'codified' in the formalized language in question), if the basic theorems can be shown to be theorems, and the (new) basic rules valid deductions in the sense of the semantic definitions in III 3, p. 61 f. This holds also if we are concerned with deductions in an extended sense, e.g. of the following kind:

If every formula in S is (generally-) valid, then also A.

Deductions of this kind are required e.g. for the foundation of rules such as

(1) From $\mathsf{A}(a)$ may be inferred $\bigwedge a \mathsf{A}(a)$;

or

(2) from $\bigvee a \mathsf{A}(a)$ and $\mathsf{A}(a) \rightarrow \mathsf{B}$, where a does not occur in B, we may infer B[2].

For the reduction of the original rules to theorems there are various possibilities, but these differ more from the point of view of interpretation than in symbolic representation. Thus in a 'purely formal' way rules of the form

(3) $$\frac{\mathsf{A}}{\mathsf{B}} \quad \text{or} \quad \frac{\mathsf{A} \quad \mathsf{B}}{\mathsf{C}}$$

may be replaced by corresponding theorems

(4) $$\mathsf{A} \rightarrow \mathsf{B} \quad \text{or} \quad \mathsf{A} \rightarrow (\mathsf{B} \rightarrow \mathsf{C})$$

and, applying the rule of inference *modus ponens*

(5) $$\frac{\mathsf{A} \quad \mathsf{A} \rightarrow \mathsf{B}}{\mathsf{B}}$$

the original rules may be re-derived from the theorems.

If the rules with two premises had been similarly replaced by a theorem $\mathsf{A} \wedge \mathsf{B} \rightarrow \mathsf{C}$, then correspondingly a rule

(6) $$\frac{\mathsf{A} \quad \mathsf{B}}{\mathsf{A} \wedge \mathsf{B}}$$

would have been required in addition.

The fact that a formula A is derivable from a set S according to rules stipulated in any way whatsoever, is often expressed by '$\mathsf{S} \vdash \mathsf{A}$', the sign

'$\vdash$' here belonging to the language in which statements are made *about* a symbolic language. Since on the basis of rules (5) and (6), it holds that $\mathsf{A}, \mathsf{A} \rightarrow \mathsf{B} \vdash \mathsf{B}$ and $\mathsf{A}, \mathsf{B} \vdash \mathsf{A} \wedge \mathsf{B}$, we may refer to these rules by '$\mathsf{A}, \mathsf{A} \rightarrow \mathsf{B} \vdash \mathsf{B}$' and '$\mathsf{A}, \mathsf{B} \vdash \mathsf{A} \wedge \mathsf{B}$', respectively. We proceed analogously in other cases.

To place a logic constructed in this way on a semantic foundation, we must therefore on the one hand validate rules (5) (and, where appropriate (6)), by the valid deductions $\mathsf{A}, \mathsf{A} \rightarrow \mathsf{B} \vDash \mathsf{B}$, (and $\mathsf{A}, \mathsf{B} \vDash \mathsf{A} \wedge \mathsf{B}$), the sign of consequence '$\vDash$' being referred in each case to the language in question. On the other hand, the formally introduced theorems (4) must be shown to be theorems in the semantic sense of III 3, p. 60. In the case of propositional logic and to a large extent[3] also in that of predicate logic, this can be done by the method outlined in III 3, p. 57. In other cases arising in predicate logic we need to make use of a number of immediate inferences, such as that every formula $\bigwedge a \mathsf{A}(a) \rightarrow \mathsf{A}(a)$ is generally-valid.

It has been shown, however, that in the case of predicate logic some rules cannot be transformed into theorems in the semantic sense exemplified in the transition from (3) to (4), and these are precisely those rules that, like (1) and (2), require an extension of the concept of consequence for their foundation. Such rules (or, at any rate, one of this type) will thus have to be retained together with (5) as basic rules. A possible form for such a system of basic theorems and basic rules will be shown in IV 2, p. 74 f.

The transition from the rules (3) to the theorems (4) may also be interpreted differently, so that the sign $\rightarrow$ expresses by definition the 'validity' of the corresponding rule and the sign combination $\mathsf{A} \rightarrow \mathsf{B}$ is *simultaneously* introduced as a formula. Rule (5) then simply expresses the fact that the transition from (3) to (4) may be reversed. Now validity attaches to precisely those rules that express the ways in which inferences may be combined. Suppose, for example, that $\mathsf{A} \rightarrow \mathsf{B}$ and $\mathsf{B} \rightarrow \mathsf{C}$ express the validity of the rules $\frac{\mathsf{A}}{\mathsf{B}}$ and $\frac{\mathsf{B}}{\mathsf{C}}$. These latter yield the compound rule

$$\cfrac{\cfrac{\mathsf{A}}{\mathsf{B}}}{\mathsf{C}}$$

and hence $\frac{\mathsf{A}}{\mathsf{C}}$, the validity of which is expressed by $\mathsf{A} \rightarrow \mathsf{C}$. Then this may be stated through the validity of the rule

(7) $$\frac{A \to B \quad B \to C}{A \to C},$$

that is, by the 'theorem'

(8) $$(A \to B) \to ((B \to C) \to (A \to C)).$$

The step from (7) to (8) is thus an example of the transition from (3) to (4). By similar methods P. Lorenzen [1] has been able to found a large part of logic on a 'constructive' basis.

The method used for the foundation of (7) and (8) is characterized by the fact that here the validity of a rule or formula is not demonstrated on the basis of a specific definition of validity, but on very weak assumptions about any validity whatsoever – assumptions that are satisfied among others by the general validity defined above in III 3. We did not there define $\to$ by the transition from (3) to (4), but this transition is contained in the relation discussed in III 3, p. 61, between $\to$ and '$\models$'. By employing special techniques, which lack of space prevents us from discussing here, it is even possible to include rules of the type of examples (1) and (2). See Quine [3], Gumin–Hermes [1].

Once the use of the sign $\to$ has been regulated in such a way, whatever the basis of proof, that we have at our disposal on the one hand the higher-order rule contained in the transition from (3) to (4) (the rule of introduction of implication, the deduction theorem[4]) viz:

> Any premise of a rule may be eliminated as rule premise by being placed as implication premise before the conclusion,

and on the other hand rule (5), then there exists a wide measure of freedom so far as the characterization of the remaining logical symbols is concerned.

The basic theorem	may be	the basic rule
$A \wedge B \to A$	exchanged	$A \wedge B \vdash A$
$A \to (B \to A \wedge B)$	for	$A, B \vdash A \wedge B$
$A \to A \vee B$		$A \vdash A \vee B$

and, to give a more complicated example,

$$(A \to C) \to ((B \to C) \to (A \vee B \to C))$$

may be replaced by

$$A \rightarrow C, B \rightarrow C, A \vee B \vdash C.$$

Thus the terms 'theorem logic' and 'rule logic' merely indicate two different aspects of logic. Theorems are in a sense 'frozen' rules and rules are 'unfrozen' theorems.

2. CALCULI

In IV 1 we considered rules of inference from two points of view: as the form in which the laws of logic may be applied, and as a means of systematizing logical laws formulated in theorems. This gives us a method for recognizing theorems as such even where the original definition of the concept of the theorem does not directly afford a procedure for doing so. We then need to refer to this definition merely in order to establish:

(1) that specific *wffs* are theorems on the basis of their structure,

and

(2) that specific formal operations, when applied to theorems, yield other theorems.

Thus, at any rate, all *wffs* that can be obtained in accordance with (1) and (2) are theorems.

Frequently, however, the method contained in (1) and (2) is freed from its attachment to a previously given concept of the theorem. A syntactic definition of the theorem is set up by stipulating:

(3) that certain *wffs* characterized by their 'grammatical' structure[5] are theorems *(basic theorems, axioms)*,

(4) that certain operations[5], applied to theorems, yield other theorems; to each of these operations there corresponds a *basic rule* which expresses that the operation produces theorems from theorems,

(5) that only what can be constructed in accordance with (3) and (4) shall be accepted as a theorem.

This schema gives rise to definitions exactly similar to those used for the definition of the formal systems or languages $\mathbf{L}_P$, $\mathbf{L}_\Sigma$, $\mathbf{L}_F$ in III 2. But in general the operations here no longer have the simple 'composing' structure of the latter, as is shown by the 'formation' of B from A and A→B when applying the rule of *modus ponens*.

There is a similar syntactic counterpart to the semantic definition of consequence, which usually involves a restriction to finite sets of premises. As symbolic representations of consequences – technically 'sequences' – sign strings are formed, consisting of

(a) a series (possibly empty) of formulas[6], these being the *premises* of the sequence (this series represents the *set* of premises of the deduction);

(b) a 'follows' *sign*, e.g. ▶, which now counts as a component of the expression, but must be distinguished from the sign for implication (here →);

(c) a formula as *conclusion*.[7]

Sequences accordingly have the form $A_1 \ldots A_n$ ▶ B, A ▶ B, or ▶ B. A syntactic 'theorem definition' for sequences must thus stipulate:

(6) that specific sequences, i.e. the *basic sequences*, are theorems;

(7) that specific operations, applied to theorems, always yield theorems (again every operation involves a *basic rule*);

(8) that theorems are obtained only by the application of (6) and (7).

The representation of *consequences* by means of *sequences* instead of rules thus enables us to systematize valid deductions without the use of higher-order rules.

We can speak of a *calculus* if there is

(A) a structurally described language, such as $\mathbf{L}_P$, $\mathbf{L}_\Sigma$, $\mathbf{L}_F$ or a totality of sequences formed from the formulas of another language;

(B) a structural theorem definition, such as (3)–(5) or (6)–(8); or else the closely related definition of derivability from a set of premises, e.g. in the form:

From any set of premises S whatsoever, may be *derived:*

(9) all basic theorems,

(10) all expressions in S,

(11) with the premises of a basic rule also its conclusion,

(12) only what can be derived in accordance with (9)–(11).

Thus, for example, all theorems are derivable from any S whatsoever. We symbolize 'S $\vdash_C$ A' for 'In the calculus *C* the expression A is derivable from S'.

This genetically described derivability involves, of course, the existence of a derivation.

By a *derivation of* A *from* S may be understood e.g.

(C) a tree-like figure (as in I 3, p. 20 f.), ending in A, with all starting

points being either basic theorems or belonging to S, and all horizontal conclusion lines indicating the application of basic rules;
or it may be introduced as

(D) a series of expressions ending in A, every member of which must be justified – either as basic theorem or as member of S or by the application of a basic rule of one or more of the members preceding it in the series.

A member of the derivation may be 'applied' any number of times.

Although this latter interpretation is sometimes less instructive, it probably corresponds more closely to the actual temporal development of a thought process.

We shall now introduce briefly a number of calculi that merit attention because of their simplicity or importance.

The Calculus of Pure Sequences (SC)

This calculus (Gentzen [1]) is designed to describe those formal properties of consequences, which are quite independent of the structure of the formulas (cf. III 3, p. 62).

The *basic signs* are: ▸ and the propositional variables $p_1, p_2, \ldots$

The *expressions* are: the sequence $p_{i_1}, \ldots, p_{i_n}$ ▸ p_k, with the limiting cases p_i ▸ p_k and ▸ p_k.

The *basic sequences* are: the sequences of the form p_i ▸ p_i

The *basic rules* are

the rule of *premise transposition*

$\ldots p_i p_j$ --- ▸$p_k \vdash_S \ldots p_j p_i$ --- ▸ p_k,

the rule of *premise fusion:*

$\ldots p_i p_i$ --- ▸ $p_k \vdash_S \ldots p_i$ --- ▸ p_k[8]

the rule of *premise expansion:*

$\ldots$ ▸ $p_k \vdash_S p_i \ldots$ ▸ p_k,

the rule of the *cut:*

$\ldots$ ▸ p_i, p_i --- ▸ $p_k \vdash_S \ldots$ --- ▸ p_k.

Clearly all the theorems of this calculus have the simple form $\ldots p_i$--- ▸ p_i. Of greater interest are the derivable rules. The calculus is primarily designed for the task of discovering for a given set of sequences one set that whilst being equivalent with regard to the mutual derivability of the sequences, is as simple as possible. Of interest, too, is the fact that it is possible, by refinements of the simple structure of SC, to arrive at calculi for propositional logic and predicate logic, which represent the peculiar relationship between 'if ... then', 'follows' and 'derivable from' better than the other calculi normally used.

The Propositional Calculus of Sequences (PSC)

Although we have introduced the sign ▶ as a symbolic counterpart to the predicate 'follows', which was defined semantically for formulas and sets of formulas, the sequences could equally well be introduced as generalized implications. But this would give us only a part of propositional logic. However, SC yields a calculus for the whole of two-valued propositional logic, if instead of propositional variables we admit all formulas of $\mathbf{L}_P$ and include additional basic theorems (or basic rules) to describe the operation of the propositional logic functors.

On the basis of IV 1, p. 69, we select the following basic rules for →:

$$\ldots \mathsf{A} \blacktriangleright \mathsf{B} \vdash_{PL} \ldots \blacktriangleright (\mathsf{A} \rightarrow \mathsf{B}) \text{ and } \ldots \blacktriangleright (\mathsf{A} \rightarrow \mathsf{B}) \vdash_{PL} \ldots \mathsf{A} \blacktriangleright \mathsf{B}$$ [9]

We then require only basic sequences for the remaining functors, thus e.g.

for ∧: $\mathsf{AB} \blacktriangleright (\mathsf{A} \wedge \mathsf{B})$, $(\mathsf{A} \wedge \mathsf{B}) \blacktriangleright \mathsf{A}$, $(\mathsf{A} \wedge \mathsf{B}) \blacktriangleright \mathsf{B}$,
for ∨: $\mathsf{A} \blacktriangleright (\mathsf{A} \vee \mathsf{B})$, $\mathsf{B} \blacktriangleright (\mathsf{A} \vee \mathsf{B})$,
$(\mathsf{A} \rightarrow \mathsf{C})(\mathsf{B} \rightarrow \mathsf{C})(\mathsf{A} \vee \mathsf{B}) \blacktriangleright \mathsf{C}$,
for ↔: $(\mathsf{A} \rightarrow \mathsf{B})(\mathsf{B} \rightarrow \mathsf{A}) \blacktriangleright (\mathsf{A} \leftrightarrow \mathsf{B})$,
$(\mathsf{A} \leftrightarrow \mathsf{B}) \blacktriangleright (\mathsf{A} \rightarrow \mathsf{B})$, $(\mathsf{A} \leftrightarrow \mathsf{B}) \blacktriangleright (\mathsf{B} \rightarrow \mathsf{A})$,
for ¬: $\mathsf{A}, \neg \mathsf{A} \blacktriangleright \mathsf{B}$, $(\mathsf{A} \rightarrow \mathsf{B})(\neg \mathsf{A} \rightarrow \mathsf{B}) \blacktriangleright \mathsf{B}$.

Again, one could introduce basic rules (instead of basic sequences), and this would be of advantage e.g. if one wished to discuss specific standard forms for proofs, for example, proofs in which all occurring formulas are sub-formulas of the end formula of the proof.[10] However, as we have introduced the sequences themselves as symbolic representations as well of consequences as of deductions (and therewith in a sense, of rules), it is probably more appropriate to represent the basic laws of propositional logic by means of basic sequences rather than by sequence rules (which in a sense are representations of *higher-order* rules).

Finally, as example of how to handle PSC, we give below a simple proof, in which the application of rules is represented by inference figures as in I.

$$\dfrac{\dfrac{\neg A \blacktriangleright (A \vee \neg A)}{\blacktriangleright (\neg A \rightarrow A \vee \neg A)} \qquad \dfrac{\dfrac{A \blacktriangleright (A \vee \neg A)}{\blacktriangleright (A \rightarrow A \vee \neg A)} \qquad (A \rightarrow A \vee \neg A)(\neg A \rightarrow A \vee \neg A) \blacktriangleright (A \vee \neg A)}{(\neg A \rightarrow A \vee \neg A) \blacktriangleright (A \vee \neg A)}}{\blacktriangleright (A \vee \neg A)}$$

As will be readily seen, all top sequences in this figure are basic sequences, and the horizontal conclusion lines indicate the application of basic rules; the 'end sequence' is thus a theorem of PSC.

The reader should now try to prove for himself the sequence $\blacktriangleright A$, which states that $\vDash_P A$ – at any rate for simple formulas A, whose general-validity he has established by the method of value-assignments explained in III 3, p. 57. The question whether such a sequence can be proved in all cases will be discussed in IV 3, p. 81 f.

The Predicate Calculus (FC)

We can obtain a predicate logic calculus from any propositional logic calculus whatsoever, by admitting *wffs* from $\mathbf{L}_F$ instead of from $\mathbf{L}_P$ and adding a number of axioms or rules by means of which the use of the functors $\wedge$ and $\vee$ or of the operators $\wedge x$ and $\vee x$ (where 'x' stands for arbitrary variables a_i), is regulated. In many cases, however, greater recognition is given to the fact that propositional logic is 'simple' in relation to predicate logic, by using as predicate logic basic theorems all *wffs* A that (in any sense whatsoever) become propositional logic theorems if all *sub-wffs* of A that are irreducible in propositional logic are replaced by propositional variables – different *sub-wffs* of A being replaced by different propositional variables.

Further, if in the process we define the propositional logic theorems according to the semantics of III 3, we obtain those formulas of predicate logic that are valid in propositional logic. This gives us a possible form for FC if we stipulate as follows:[11]

the *basic theorems* are all F-*wffs* valid in propositional logic;

the *basic rules* (formulated with '... $\vdash_F$ ...' for 'from ... is derivable ...') are:

modus ponens: A, $(A \rightarrow B) \vdash_F B$,[12]

the $\wedge$-*rules:* (to be called *subsequent generalization* Gs)

$A \rightarrow B \vdash_F A \rightarrow \wedge x B$, in the case where x does not occur in A,

and (to be called *initial generalization* Gi)

$A \rightarrow B \vdash_F \wedge x A \rightarrow B$, (without restriction),

the $\vee$-*rules:* (to be called *subsequent particularization* Ps)

$A \rightarrow B \vdash_F A \rightarrow \vee x B$ (without restriction),

and (to be called *initial particularization* Pi)

$A \rightarrow B \vdash_F \vee x A \rightarrow B$, in the case where x does not occur in B;

the rules for the substitution of variables:

(to be called *bound re-naming* Rb)

$A \vdash_F A^*$, in the case where A^* is obtained from A by the simultaneous substitution of another variable not occurring in B for a '$\wedge$-variable' or '$\vee$-variable' x occurring in A in $\wedge x$ or $\vee x$ and in the subsequent subformula B of A – the *scope* of $\wedge x$ or $\vee x$;

and (to be called *rule for term substitution* TS)

$A \vdash_F A^*$, in the case where A^* is obtained from A by the substitution of one and the same term t for a specific variable x at all positions of A where x does not occur in $\wedge x$ or $\vee x$ or in the scope of $\wedge x$ or $\vee x$. In the process, however, no variable y occurring in t may come within the scope of a $\wedge y$ or $\vee y$ that is, no x that is to be replaced may occur within the scope of such an operator. If these conditions are satisfied, the application of TS may also be symbolized by '$A \vdash_F A(x/t)$', and frequently '$A(x/t)$' is meant to indicate that all impediments to the substitution have previously been removed in A by means of bound re-namings.

These basic theorems and basic rules have been chosen so that the theorems yielded by them are generally-valid formulas precisely in the sense of III 3, p. 60. This aspect will be discussed, at any rate in principle, in IV 3. The reader should note that in III 2, C(10), p. 53, in contrast to our procedure in II 2, we use $\wedge x$ ... and $\vee x$ ... in the customary sense as basic concepts. (Thus account is taken of x ... or resp. [x] ... only in the combinations $\wedge$[x] ... and $\vee$[x] .... On the general use of [x] ... cf. V 2, p. 94 f.).

Lack of space prevents us from citing even the most important theorems with their proofs. Rather than present the reader with more theorems without proofs, we shall clarify the use of the rules by means of some simple examples of their application.

The four rules for $\wedge$ and $\vee$ have been chosen so as to bring out most clearly their common and disparate features. The properties of $\wedge$ and $\vee$ expressed in the simpler basic rules Gi and Ps may also be formulated in theorems. The very simple proofs of these theorems are as follows (where 'PL' stands for 'valid in propositional logic'):

$$\frac{A \to A \quad \text{(PL)}}{\wedge x A \to A} \qquad \text{and} \qquad \frac{B \to B \quad \text{(PL)}}{B \to \vee x B}\text{.}$$

Inversely, however, these theorems may also be used as basic theorems to replace the corresponding basic rules, which then become derivable rules. The derivation for Gi is:

$$\dfrac{\underset{A \to B}{\textit{(rule premise)}} \qquad \dfrac{\underset{\wedge xA \to A}{\textit{(possible axiom)}} \quad \underset{(\wedge xA \to A) \to (A \to B \;\cdot\!\to\!\cdot\; \wedge xA \to B)}{\text{(PL)}}}{A \to B \;\cdot\!\to\!\cdot\; \wedge xA \to B}}{\wedge xA \to B}$$

The derivation for Ps is entirely analogous.

With the aid of TS the theorems $\wedge xA \to A$ and $A \to \vee xA$ may be generalized into $\wedge xA \to A(x/t)$ and $A(x/t) \to \vee xA$. Often, too, the basic rules Gi and Ps are expressed in their corresponding general form; in this case TS is demonstrable.

The rule Gs is a generalization of the rule $B \vdash_F \wedge xB$, cf. IV 1, p. 67, which at first sight appears a more obvious choice. However, if this had been selected as basic rule, then a theorem of $\wedge$-transference, i.e. formulas of the form $\wedge x(A \to B) \to (A \to \wedge xB)$, which are always generally-valid if x does not occur in A, would not be demonstrable for the general case.[13] We demonstrate first the above-mentioned simpler rule of generalization:

Let A be a formula in which x does not occur.

$$\dfrac{\underset{A \to A}{\text{(PL)}} \qquad \dfrac{\dfrac{\underset{B}{\textit{(rule premise)}} \quad \underset{B \;\cdot\!\to\!\cdot\; (A \to A) \to B}{\text{(PL)}}}{(A \to A) \to B}}{(A \to A) \to \wedge xB}\;\text{(Gs)}}{\wedge xB}$$

The above-mentioned theorem of $\bigwedge$-transference may be proved as follows:

$$\dfrac{\dfrac{\dfrac{\dfrac{\dfrac{\dfrac{\overset{(PL)}{(A \to B) \to (A \to B)}}{\bigwedge x(A \to B) \to (A \to B)} \qquad \overset{(PL)}{\left.\begin{matrix}\bigwedge x(A \to B) \to (A \to B) \\ \cdot\!\to\!\cdot \;\; \bigwedge x(A \to B) \wedge A \to B\end{matrix}\right\}}}{\bigwedge x(A \to B) \wedge A \to B}\,{}^{14}}{\bigwedge z(A \to B^*) \wedge A \to B}\,(Gs)}{\bigwedge z(A \to B^*) \wedge A \to \bigwedge x B}\,{}^{15}}{\bigwedge x(A \to B) \wedge A \to \bigwedge x B} \qquad \overset{(PL)}{\left\{\begin{matrix}\bigwedge x(A \to B) \wedge A \to \bigwedge x B \\ \cdot\!\to\!\cdot \\ \bigwedge x(A \to B) \to (A \to \bigwedge x B)\end{matrix}\right\}}}{\bigwedge x(A \to B) \to (A \to \bigwedge x B)}$$

Inversely, Gs may be obtained again from the simple rule of generalization with the theorem of $\bigwedge$-transference as additional axiom:

$$\dfrac{\dfrac{\overset{\textit{(rule premise)}}{A \to B}}{\bigwedge x(A \to B)} \qquad \overset{\textit{(possible axiom)}}{\bigwedge x(A \to B) \to (A \to \bigwedge x B)}}{A \to \bigwedge x B}$$

The basic rule Pi is a refinement of the derivable rule

$$\bigwedge x A,\ A \to B \vdash_F B, \text{ in the case where } x \text{ does not occur in } B$$

(cf. IV 1, p. 67), which is obtained from Pi by a simple application of *modus ponens*

$$\dfrac{\overset{\textit{(rule premise)}}{\bigvee x A} \qquad \dfrac{\overset{\textit{(rule premise)}}{A \to B}}{\bigvee x A \to B}}{B}$$

This derivable rule alone does not suffice to prove e.g. (generally-valid) formulas of the form $\bigvee x B \to \bigvee x(A \to B)$. But the following generalization of the rule, again derivable with Pi:

$$C \to \bigvee x A, \quad A \to B \vdash_F C \to B,$$

in the case where x does not occur in B, is equivalent to Pi: we merely substitute $\bigvee x A$ for C.

Having worked through these examples, the reader should try to derive the rules in which the syllogisms (I 2, p. 15f.) may be expressed on the basis of I 3, p. 25.

3. SOUNDNESS AND COMPLETENESS OF CALCULI

Every calculus may be manipulated as a kind of combinatorial game – merely to discover what happens – and in the process one can learn a great deal about the connexions between the 'theorems' of the calculus. In general, however, we tend to be most interested in those calculi whose basic theorems and basic rules we recognize in some sense or other. This may express the following intention: whatever the signs or expressions in the calculus may be capable of meaning, we shall consider only those interpretations where the selected basic theorems and basic rules hold. Once we have agreed on these, we must also recognise all demonstrable theorems, since the correct application of the rules can be controlled. Theories in which modalities (such as *necessary*, *possible*) occur as definable or as basic concepts have usually been presented in this form. Cf. Lewis-Langford [1]. More recently S. Kripke has put forward a semantic approach to modality which raises similar questions of soundness and completeness of related calculi. See Schütte [1].

Here the totality of admissible interpretations is in a sense defined precisely by the choice of calculus; but it is not stated explicitly, since the language to be interpreted is used only within the range determined by the calculus selected. In the case of the calculi discussed in IV 2, for example, certain basic theorems or basic rules, whose foundation presupposes the notion of a closed domain of objects, must be omitted or replaced by weaker ones, should one consider this notion to be untenable when referred to infinite domains.

A closer analysis shows that such weakening needs to be undertaken already in propositional logic and that in particular finite-valued matrices are no longer adequate to represent propositional connexions, although the basic rules of predicate logic may be retained. By far the most important among the variants proposed is the so-called intuitionist propositional and hence also predicate logic.[16] This may be regarded as the totality of theorems and rules that hold independently of the assumption of closed infinite domains of objects, but is often defined by means of calculi, whose basic theorems and basic rules are at any rate compatible with this critical standpoint. We may obtain a calculus for intuitionist propositional logic e.g. from PSC (p. 73 f.) by replacing the basic sequence $(A \to B)\ (\neg A \to B) \blacktriangleright B$ by the sequence $(A \to B)(A \to \neg B) \blacktriangleright \neg A$, which is demonstrable in PSC. In this calculus we cannot derive the sequences $\blacktriangleright (A \vee \neg A)$ and $\neg\neg A \blacktriangleright A$, which express the assumption that every *proposition* is either *true* or *false*[17], an assumption also underlying the discrete ontology introduced in II 1, p. 28.

If, on the other hand, we start with the concepts of validity and of consequence for propositional and predicate logic, whose definitions presuppose this ontology, and if we regard the calculi as aids for determining that e.g. $\vDash_P A$ or respectively that $S \vDash_F A$, then the calculi, to be 'usable', must satisfy certain conditions.

For the formulation of such presuppositions let '$\vDash_L$' stand for validity[18] or equally for the consequence relation in reference to a given language **L**, and '$\vdash_C$' for demonstrability or equally for derivability from a set of premises in the calculus *C*.

If *C* satisfies the following condition with reference to **L**:

(1) If $\vdash_C A$, then $\vDash_L A$,

then *C* is usable for the discovery of valid formulas in **L**. In this sense, for example, the FC discussed in IV 2, p. 74 f, is usable for predicate logic.

Sometimes the following requirement is made in addition to (1):

(2) If $A_1, \ldots, A_n \vdash_C B$, then $A_1, \ldots, A_n \vDash_L B$;

that is, *C* is intended to be usable also for the discovery of consequences. The requirement is not met by the FC discussed in IV 2, for three of its basic rules (viz: Gs, Pi and TS) infringe it. For example, under TS $A^1a_1 \vdash_F A^1a_2$ holds but not $A^1a_1 \vDash_F A^1a_2$; for in this case, by reason of III 3 (1), p. 61, the formula $A^1a_1 \to A^1a_2$ would be generally-valid, which is easily disproved. In fact, only special cases or else alterations of (2) can be demonstrated for the FC in IV 2, such as, for example:[19]

(2.1) If $\vdash_C A_1 \wedge \ldots \wedge A_n \to B$, then $A_1, \ldots, A_n \vDash_L B$.

However, it is also possible to design calculi which will allow consequences to be directly discovered. Among these are sequence calculi, where 'usability' may be formulated as follows:

(3) If $\vdash_C \mathsf{A}_1 \ldots \mathsf{A}_n \blacktriangleright \mathsf{B}$, then $\mathsf{A}_1, \ldots, \mathsf{A}_n \vDash_L \mathsf{B}$.

This formulation expresses the close connexion on the one hand between $\blacktriangleright$ and '$\vDash$', and on the other, in virtue of its similarity to (2.1), between $\blacktriangleright$ and $\rightarrow$. In fact, it is easy to design a calculus *C* for propositional logic such that (2) holds. Our intention in presenting the PSC in IV 2, p. 73 f, was to give a simple example of a calculus having the property (3). This calculus can furthermore be easily converted into a calculus FSC for predicate logic characterized by the property (3).

The property of 'usability' of a calculus, which is expressed by (1), (2) or (3) merely means that a calculus thus characterized will produce *no false derivations*. Let us designate this quality somewhat more cautiously as *soundness* (with reference to a given concept of validity or consequence). Proofs for the soundness of calculi expressed in the form of (1) or (3) all have the same pattern: it is shown that the basic theorems are sound and that the application of the basic rules cannot produce unsound conclusions from sound premises. In demonstrating soundness as formulated in (2), one must bear in mind that the basic theorems are to be manipulated like basic rules without premises, and make use of the fact that consequences may be put together like derivations.

If a calculus *C* is to be truly usable with respect to **L**, then apart from being sound it must produce validity or consequence for **L** in a sufficient number of cases, if possible in all. In this case *C* is said to be *complete* in respect of **L**. This completeness, which in general makes sense only for sound calculi, is expressed by the conversions of (1), (2) or respectively (3), viz:

(4) If $\vDash_L \mathsf{A}$, then $\vdash_C \mathsf{A}$,
(5) If $\mathsf{A}_1, \ldots, \mathsf{A}_n \vDash_L \mathsf{B}$, then $\mathsf{A}_1, \ldots, \mathsf{A}_n \vdash_C \mathsf{B}$,
(6) If $\mathsf{A}_1, \ldots, \mathsf{A}_n \vDash_L \mathsf{B}$, then $\vdash_C \mathsf{A}_1 \ldots \mathsf{A}_n \blacktriangleright \mathsf{B}$.

The PSC in IV 2, p. 73 f. is complete in the sense of (6) for propositional logic, and the FC in IV 2, p. 74, is complete in the sense of (4) for predicate logic. Calculi are also known which are complete for predicate logic in the sense of (5) or alternatively (6).

It is in general more difficult to prove the completeness of a calculus than to prove its soundness (this latter is sometimes proved incidentally in the course of proving the former). Proofs for the completeness of most (complete) calculi for propositional logic are relatively easy. The reason for this is that in this case the definition of general validity yields a method of proof and hence a calculus, though not in the sense of our standardization. In order to adapt this method to our standard form e.g. in the case of PSC, we proceed as follows: If A is a formula containing (for example) precisely the variables $p_1, \ldots, p_n$, then the evaluation of A with the aid of truth tables is reflected in 2^n *demonstrable* sequences of the form

$$(7)\qquad \begin{Bmatrix} p_1 \\ \neg p_1 \end{Bmatrix} \cdots \begin{Bmatrix} p_k \\ \neg p_k \end{Bmatrix} \blacktriangleright \begin{Bmatrix} \mathsf{A} \\ \neg \mathsf{A} \end{Bmatrix}$$

where to the left of ▶ are entered all 2^n possible value assignments for $p_1, \ldots, p_n$ ('p_i' for 'p_i is true' and '$\neg p_i$' for 'p_i is false'), and to the right of ▶ we write A or $\neg$A according to the value of A for the corresponding value assignment on the left. The fact that for any A all these sequences are provable is demonstrated in the first part of the proof. This is done step by step *via* the construction of A. *If* A is generally valid, then A alone occurs at all positions on the right. In this case it is possible to demonstrate – essentially by applying the so called 'deduction theorem' $\ldots \mathsf{A} \blacktriangleright \mathsf{B} \vdash_L \ldots \blacktriangleright \mathsf{A}\rightarrow\mathsf{B}$ – the 2^{n-1} pairs of sequences:

$$(8a)\qquad \begin{Bmatrix} p_1 \\ \neg p_1 \end{Bmatrix} \cdots \begin{Bmatrix} p_{n-1} \\ \neg p_{n-1} \end{Bmatrix} \blacktriangleright p_n \rightarrow \mathsf{A},$$

$$(8b)\qquad \begin{Bmatrix} p_1 \\ \neg p_1 \end{Bmatrix} \cdots \begin{Bmatrix} p_{n-1} \\ \neg p_{n-1} \end{Bmatrix} \blacktriangleright \neg p_n \rightarrow \mathsf{A}.$$

By the application of a basic sequence of the form $(\mathsf{B}\rightarrow\mathsf{A})\ (\neg\mathsf{B}\rightarrow\mathsf{A}) \blacktriangleright \mathsf{A}$, viz: $(p_n\rightarrow\mathsf{A})\ (\neg p_n\rightarrow\mathsf{A}) \blacktriangleright \mathsf{A}$, we then obtain the 2^{n-1} sequences

$$(9)\qquad \begin{Bmatrix} p_1 \\ \neg p_1 \end{Bmatrix} \cdots \begin{Bmatrix} p_{n-1} \\ \neg p_{n-1} \end{Bmatrix} \blacktriangleright \mathsf{A}.$$

By repeated application of this process all premises are eliminated and – for a generally-valid A – we obtain the demonstrable sequence ▶ A. Although this is only a special case of (6), it indicates a generally applicable method.

In the case of predicate logic, the definition of general validity would seem to yield no procedure, and the considerations on which is based the soundness of FC as an already standardized method of proof, do not suffice to demonstrate completeness. On the contrary, it would seem that there will always be non-derivable rules that can be shown to be admissible by intuitive means.

We should now like to outline the basic idea of one of the more recent proofs of completeness.[20] If $\vDash_F$ A, then A is *valid* in particular for the somewhat artificial domain **D** consisting of all terms (i.e. of specific sign strings) and in respect of that fixed value assignment for a_i and f_k^i where every term (as part of the formula) denotes itself (as member of the domain), i.e. $\mathfrak{B}^*(\mathsf{t}) = \mathsf{t}$. That there is such a value-assignment for f_k^i must, of course, be demonstrated; but this is a simple matter. Next we must deal with the value-assignment for the A_k^i variables. Now the **D**-*validity* defined by such value-assignments is such that $\mathfrak{B}^*(\mathsf{A}) = \mathbf{T}$ for all those value-assignments which satisfy, apart from the propositional logic conditions, also the conditions

(10) $\mathfrak{B}^*(\bigwedge x\mathsf{C}) = \mathbf{T}$ if and only if for all t in **D** $\mathfrak{B}^*(\mathsf{C}(x/\mathsf{t}))=\mathbf{T}$

and

(11) $\mathfrak{B}^*(\bigvee x\mathsf{D}) = \mathbf{T}$ if and only if for at least one t in **D** $\mathfrak{B}^*(\mathsf{D}(x/\mathsf{t})) = \mathbf{T}$,

for all formulas[21] beginning with $\bigwedge$ or $\bigvee$. If there were only the two formulas $\bigwedge x\mathsf{C}$ and $\bigvee x\mathsf{D}$ to be taken into account, and if the only terms were the variables a_1 and a_2, then this could be expressed by the fact that A follows from[22]

(12) $\bigwedge x\mathsf{C} \leftrightarrow \mathsf{C}(x/a_1) \wedge \mathsf{C}(x/a_2)$

and

$\bigvee x\mathsf{D} \leftrightarrow \mathsf{D}(x/a_1) \vee \mathsf{D}(x/a_2)$

already on the basis of propositional logic.

According to the laws of propositional logic an equivalent transformation for this is provided by the fact that on the basis of propositional logic A follows from each of the four sets of the form

(13) $$\mathsf{S}_{i,j} = \{\bigwedge x\mathsf{C} \to \mathsf{C}(x/a_1)\quad \bigwedge x\mathsf{C} \to \mathsf{C}(x/a_2),\; \mathsf{D}(x/a_1) \to \bigvee x\mathsf{D},\quad \mathsf{D}(x/a_2) \to \bigvee x\mathsf{D},\; \mathsf{C}(x/a_i) \to \bigwedge x\mathsf{C},\quad \bigvee x\mathsf{D} \to \mathsf{D}(x/a_j)\},$$

where i and j stand independently of each other for the values 1 and 2. A choice appropriate to the following must now be made among these four sets. Let us suppose that $i=1$, $j=2$ is an 'appropriate' choice (see below). Then the following formula obtained from $\mathsf{S}_{1,2}$ by transformation of the premises of deduction into premises of implication (by application of the deduction theorem according to IV 1, p. 69), is valid already on the basis of propositional logic (and is thus an axiom of FC):

(14) $$(\bigwedge x\mathsf{C} \to \mathsf{C}(x/a_1)) \to (\ldots \to ((\mathsf{C}(x/a_1) \to \bigwedge x\mathsf{C}) \to ((\bigvee x\mathsf{D} \to \mathsf{D}(x/a_2)) \to \mathsf{A}))\ldots).$$

The first four premises are demonstrable in FC by means of Gi, Ps and TS, and may therefore be 'cut' by use of *modus ponens*.[23] We have thus proved within FC the formula

(15) $$(\mathsf{C}(x/a_1) \to \bigwedge x\mathsf{C}) \to ((\bigvee x\mathsf{D} \to \mathsf{D}(x/a_2)) \to \mathsf{A}).$$

By means of propositional logic transformations[24] we now derive the two formulas

(16) $$\bigwedge x\mathsf{C} \to ((\bigvee x\mathsf{D} \to \mathsf{D}(x/a_2)) \to \mathsf{A}),$$

(17) $$\neg\, \mathsf{C}(x/a_1) \to ((\bigvee x\mathsf{D} \to \mathsf{D}(x/a_2)) \to \mathsf{A}).$$

From formula (17), and using the rule derivable from Gs and TS, viz:

(18) $$\neg\, \mathsf{C}(x/a) \to \mathsf{B} \vdash_{\mathrm{F}} \neg \bigwedge x\mathsf{C} \to \mathsf{B}$$
(in the case where a does not occur in $\mathsf{C} \to \mathsf{B}$)

we obtain a proof for

(19) $$\neg \bigwedge x\mathsf{C} \to ((\bigvee x\mathsf{D} \to \mathsf{D}(x/a_2)) \to \mathsf{A})$$

and from (16) and (19), again on the basis of propositional logic:[25]

(20) $$(\bigvee x\mathsf{D} \to \mathsf{D}(x/a_2)) \to \mathsf{A}.$$

This gives us, as above by rules of propositional logic, two formulas:

(21) $$\neg \bigvee x\mathsf{D} \to \mathsf{A},$$

and

(22) $\mathsf{D}(x/a_2) \to \mathsf{A}$, whence *via* Pi:
(23) $\bigvee x\mathsf{D} \to \mathsf{A}$,

and finally, again by rules of propositional logic, from (21) and (23):

(24) A.

This arrangement has been chosen so as to point out as closely as possible the analogy to the case of sentential logic. A technically simpler way is given by the possibility of using the non-occurrence of the respective variables for proving

(25) $\bigvee a_1(\mathsf{C}(x/a_1) \to \bigwedge x\mathsf{C}) \to (20)$, from (15)

and (later)

(26) $\bigvee a_2(\bigvee x\mathsf{D} \to \mathsf{D}(x/a_2)) \to \mathsf{A}$, from (20)

where the exhibited premises are provable in FC.

Under each of these arrangements, generally, the earlier applications of *modus ponens* are needed to 'free' some variable, i.e. to satisfy the non-occurrence conditions as required for (18), (23) or (25), (26) respectively. The real difficulty of the general case is that these conditions cannot be fulfilled by a previously delimited number of variables. In actual fact infinitely many variables must be introduced for the general case and a 'suitable' selection and order of sequence must be laid down for them. If there are terms (other than variables) in the language, the 'appropriate choice' includes that those places as taken by a_1 and a_2 in (15) are reserved for variables. The number of premises thus becomes infinite, and we require a special auxiliary theorem to enable us to return to a finite set of premises after a 'suitable' selection. The characteristic of the concept of deduction mentioned in III 3, p. 62, viz:

If A follows from S, then A follows from suitable finite sub-set of S, which we have here formulated for predicate logic formulas 'by rules of propositional logic', allows the transition to a finite set of premises which, as shown in (13) to (24), may then be manipulated and eliminated by the use of a suitable sequence of the variables.

NOTES

1 Cf. for example I 3, p. 19 ff., where the syllogisms are reduced to *barbara* and *darii*

and certain auxiliary modes of inference by means of higher-order rules which in essence express the structure of inferences.

[2] This inference is a frequent one in mathematical practice, when from the existence of an *a* such that A(*a*), is inferred the existence of a *b* such that B(*b*). We take as premise $\bigvee a A(a)$. Then let a_1 be 'such an *a*'. From a_1 is constructed a b_1 such that $B(b_1)$. We then infer $\bigvee bB(b)$, where no further reference is made to a_1.

[3] The reason being that in predicate logic everything holds that is 'already valid on the basis of propositional logic'.

[4] The term 'deduction theorem' reflects a situation, where this rule is not basic but a non-trivial theorem *about* a calculus in the sense of IV 2.

[5] For requirements regarding the *set* of the basic theorems and the *relations* underlying the operations, cf. VII 1, p. 123.

[6] If all outside parentheses of these formulas are written down, the formulas can be simply juxtaposed into a sign string. For the sake of legibility, however, they are usually separated by commas.

[7] If, after Gentzen [2], pp. 81 ff, the succedent is also admitted to be a series of wffs, such sequences admit a much more elegant treatment.

[8] These two rules describe the series of premises as representing a set of premises: significance attaches to neither the arrangement nor the frequency of the members of the series.

[9] This rule could also be replaced by a basic sequence, viz: $A\ (A \rightarrow B) \blacktriangleright B$; however, the rule we have selected gives greater prominence to the feature of reversibility.

[10] If one wished to carry through this idea, which has an important bearing on richer languages, one would once again have to generalize the concept of a sequence. Cf. in this connexion e.g. Scholz-Hasenjaeger [1], p. 261 ff.

[11] A detailed treatment of this form of FC will be found in Scholz-Hasenjaeger [1].

[12] This rule of course yields nothing new so long as it is applied only to basic theorems. The situation changes, however, when at least one of the other basic rules is applied.

[13] They are, however, demonstrable if e.g. $\bigwedge xB$ is demonstrable.

[14] A variable that occurs neither in A nor in B is chosen for *z*. Since, as previously stipulated, *x* does not occur in A, the substitution of *z* for *x* (1) does not alter A, (2) changes B into B* in the sense of Rb.

[15] This reverse re-naming restores the earlier formula. The insertion of the two Rb's allows a freer use of Gs (and analogously for Pi). Gs and Pi are often used in this extended sense from the start, but in this case the formulation of the conditions of applicability becomes more complex.

[16] cf. in this connexion Heyting [1], VII and Kleene [2], § 13.

[17] At any rate, this is part of what is assumed: for it is possible to construct generalized truth-tables **T** with more than two 'truth-values' where nonetheless $\vDash_T A \vee \neg A$ and $\neg\neg A \vDash_T A$ hold. (We define '$\vDash_T$' analogously to $\vDash_p$ for **T**.)

[18] That is, in most cases, general validity in the sense of the definition in III 3, p. 60. Sometimes, however, validity is defined for a narrower range of interpretations in an analogous sense, as e.g. in III 3, p. 61.

[19] We cite only the simplest variant of (2). In important cases special assumptions relating to the variables occurring in $A_1, \ldots, A_n$ have to be made, but their discussion would take us too far from our main topic.

[20] After a proof by Beth [1], p. 263.

[21] It would be enough to stipulate: for such sub-formulas of A and the formulas obtainable therefrom through TS.

[22] In fact bound re-namings are generally required here so as to allow all substitutions x/a_i. Because of this certain refinements in the basic conception become necessary.
[23] In a sequential-logic version of this proof, this step would be an application of the *cut*-rules.
[24] i.e. essentially the propositional logic theorem $((p \to q) \to r) \to (\neg p \to r) \wedge (q \to r)$.
[25] We using essentially the propositional logic theorem $(p \to q) \to ((\neg p \to q) \to q)$; cf. the application of the sequence (B → A) (¬ B → A) ▶ A on p. 81.

CHAPTER V

RICHER LOGICAL SYSTEMS

Although the language $\mathbf{L_F}$ of predicate logic as sketched in III 2 C, p. 52 f., is fairly generally applicable – i.e. to every 'world' that can be described in terms of a discrete ontology (II 1, p. 28), nevertheless it is often expedient to extend its expressive range for specific applications. We shall indicate below some of the most important of these extensions.

1. IDENTITY AND THE DEFINITE ARTICLE

A. *Identity*

The relation that holds between an object and itself and no other object, is one 2-place attribute among many others. It is, however, distinguished among these in that it is meaningful for every domain of objects, and for this reason the theory of identity is usually regarded as a part of logic. Typographically, too, we distinguish a special *identity symbol* (usually $=$) from other predicates, and the *equations* $t_1 = t_2$ from other atomic formulas. (The form $t_1 = t_2$ is more usual than $= t_1 t_2$, the full form being $(t_1 = t_2)$ together with rules which in most cases allow the parentheses to be omitted.) These new atomic formulas can be used as additional 'building bricks' in the construction of formulas. In this way $\mathbf{L_F}$ can be extended into the language $\mathbf{L_I}$ of predicate logic with identity. As is customary, we abbreviate $\neg t_1 = t_2$ to $t_1 \neq t_2$, and we stipulate that the symbol $=$ stands in all cases for the attribute of identity. This establishes general validity and consequence for the logic of identity, as symbolized by $\models_I$.

The following additional basic theorems (the axioms of identity) give a syntactic description of the concept of identity as complete as syntax can be; in other words, the formulas that can be demonstrated with the additional use of these axioms are precisely the generally valid formulas of the logic of identity. The axioms of identity are:

the so-called properties of a relation of *equivalence*, formulated for $=$,

(1) $\bigwedge x(x = x), \quad \bigwedge xy(x = y \to y = x),$
$\bigwedge xyz(x = y \wedge y = z \to x = z),$

and for arbitrary n-place predicates A^n,

(2) $\bigwedge x_1 \ldots x_n y_1 \ldots y_n(x_1 = y_1 \wedge \ldots \wedge x_n = y_n \to (A^n x_1 \ldots x_n \to A^n y_1 \ldots y_n)),$

as well as for arbitrary n-place functional variables f^n,

(3) $\bigwedge x_1 \ldots x_n y_1 \ldots y_n(x_1 = y_1 \wedge \ldots \wedge x_n = y_n \to f^n x_1 \ldots x_n = f^n y_1 \ldots y_n).$

The properties of $=$ expressed in (2) and (3) above are also called properties of *congruence*.

If the extensions of FC indicated above are standardized in the sense of IV 2, we obtain a calculus IC for predicate logic with identity, having the relation of derivability $\vdash_I$.

The following are typical theorems of IC, i.e. derivable from these axioms:

(4) $Ax \leftrightarrow \bigwedge y(y = x \to Ay),$
(5) $Ax \leftrightarrow \bigvee y(y = x \wedge Ay),$
(6) $Ax_1 \wedge \ldots \wedge Ax_n \leftrightarrow \bigwedge y(y = x_1 \vee \ldots \vee y = x_n \to Ay),$
(7) $Ax_1 \vee \ldots \vee Ax_n \leftrightarrow \bigvee y((y = x_1 \vee \ldots \vee y = x_n) \wedge Ay).$

Of these, (6) states that x_1, *and* … *and* x_n have the property A if and only if every y that is identical with x_1 *or* … *or* x_n has the property A. We leave it to the reader to formulate the remaining sentences in natural language.

Probably the most important of the additional possibilities of expression afforded by the introduction of identity, is that of rendering the 'naive' use of number words (i.e. in phrases such as '*three* cats' '*nine* bowls', …, as distinct from the abstract use[1] as in '*four* is a square number', '*three* plus *four* is *seven*' …).

Thus with the aid of identity we can express the so-called *numerical propositions*, viz:

(8) 'There are *(at least, at most, exactly)* two (three, four, …) A-things',
(9) '*(At least, at most, exactly)* two (three, four, …) A-things are B-things',

as well as the limiting cases:

(10) 'There is *(at least, at most, exactly)* one *A*-thing',
(11) '*(At least, at most, exactly)* one *A*-thing is a *B*-thing'.

In this connexion it is to be noted that the specifying expressions 'at least', 'at most', 'exactly', which are often absent from natural-language examples (in cases where they can be inferred from the context) must, where necessary, be supplied before translation into the symbolic language. Thus: 'Competitors are allowed to make two attempts' means, of course, '... at most two ...'

We give below the symbolical forms of the numerical propositions.

For (10):

$\bigvee xAx$	There is *at least* one *A*-thing.[2]
$\bigvee x\bigwedge y(Ay \rightarrow x = y)$	There is *at most* one *A*-thing.
$\bigvee x\bigwedge y(Ay \leftrightarrow x = y)$	There is *exactly* one *A*-thing.

For (11):

$\bigvee x(Ax \wedge Bx)$	*At least* one *A*-thing is a *B*-thing.[3]
$\bigvee x\bigwedge y(Ay \wedge By \rightarrow x = y)$	*At most* one *A*-thing is a *B*-thing.
$\bigvee x\bigwedge y(Ay \wedge By \leftrightarrow x = y)$	*Exactly* one *A*-thing is a *B*-thing.

In the general case the length of the formula increases so rapidly with the quantity to be described that there is little point in writing out any but the simplest cases; we shall therefore restrict our examples to the cases 'two' and 'three'.

Thus for (8):

$$\bigvee xy(x \neq y \wedge Ax \wedge Ay),$$
$$\bigvee xyz(x \neq y \wedge x \neq z \wedge y \neq z \wedge Ax \wedge Ay \wedge Az):$$

there are *at least* two (resp. three) *A*-things.

$$\bigvee xy\bigwedge z(Az \rightarrow z = x \vee z = y),$$
$$\bigvee xyz\bigwedge u(Au \rightarrow u = x \vee u = y \vee u = z):$$

there are *at most* two (resp. three) *A*-things.

$$\bigvee xy(x \neq y \wedge \bigwedge z(Az \leftrightarrow z = x \vee z = y)),$$
$$\bigvee xyz(x \neq y \wedge x \neq z \wedge y \neq z \wedge \bigwedge u(Au \leftrightarrow u = x \vee u = y \vee u = z)):$$

there are *exactly* two (resp. three) *A*-things.

The reader may not find it easy to see that the above formulas really do have the stated meanings.[4] However, once he has made the effort, he will be able to work out for himself the formulas for four, five, etc.

Lastly, for (9):

$$\bigvee xy(x \neq y \wedge Ax \wedge Bx \wedge Ay \wedge By),$$
$$\bigvee xyz(x \neq y \wedge x \neq z \wedge y \neq z \wedge Ax \wedge Bx \wedge Ay \wedge By \wedge Az \wedge Bz):$$

at least two (resp. three) *A*-things are *B*-things.

$$\bigvee xy \bigwedge z(Az \wedge Bz \rightarrow z = x \vee z = y),$$
$$\bigvee xyz \bigwedge u(Au \wedge Bu \rightarrow u = x \vee u = y \vee u = z):$$

at most two (resp. three) *A*-things are *B*-things.

$$\bigvee xy(x \neq y \wedge \bigwedge z(Az \wedge Bz \leftrightarrow z = x \vee z = y)),$$
$$\bigvee xyz(x \neq y \wedge x \neq z \wedge y \neq z$$
$$\wedge \bigwedge u(Au \wedge Bu \leftrightarrow u = x \vee u = y \vee u = z)):$$

exactly two (resp. three) *A*-things are *B*-things.

Again, the reader will have to make a mental effort to understand the formulas.[5]

The various numerical propositions – and similarly, of course, their symbolical representations – are inter-related. We give below the most important of these inter-relationships, at first in natural language.

(12) There are exactly n *A*-things if and only if there are at least n *A*-things and at most n *A*-things.

(13) There are at most n *A*-things if and only if there are not at least $n+1$ *A*-things.

(14) If there are at most n *A*-things, then there are at most $n+1$ *A*-things.

Using the symbols previously introduced, we can now re-formulate (12), (13), (14) for a fixed n (e.g. $n=1$, $n=2$, $n=3$, ...). For $n=2$ we obtain the formulas

(12.1) $$\bigvee xy(x \neq y \wedge \bigwedge z(Az \leftrightarrow z = x \vee z = y))$$
$$\leftrightarrow \bigvee xy(x \neq y \wedge Ax \wedge Ay) \wedge \bigvee xy \bigwedge z(Az \rightarrow z = x \vee z = y),$$

$$(13.1)\quad \bigvee xy \bigwedge z(Az \to z = x \vee z = y) \leftrightarrow \neg \bigvee xyz(x \neq y \wedge x \neq z \wedge y \neq z \wedge Ax \wedge Ay \wedge Az),$$

or in a formulation equivalent under the rules of FC applied to $\mathbf{L_I}$

$$(13.2)\quad \bigvee xy \bigwedge z(Az \to z = x \vee z = y) \leftrightarrow \bigwedge xyz(Ax \wedge Ay \wedge Az \to x = y \vee x = z \vee y = z).$$

This gives us another way of expressing that there are at most two (or analogously three, ...) A-things.[6]

$$(14.1)\quad \bigvee xy \bigwedge z(Az \to z = x \vee z = y) \to \bigvee xyz \bigwedge u(Au \to u = x \vee u = y \vee u = z).$$

To derive these formulas from the axioms of identity would lead us too far from our present topic.

B. *The definite article (individual description)*

Let us now try to symbolize propositions such as 'Elizabeth is the present Queen of England', 'Dickens is the author of David Copperfield', '2 is the even prime number', i.e. propositions of the form 'y is the (only) A-thing'. We require a formula of the form $y = \ldots$. With the symbols so far available to us, however, all we can manage is something along the lines of

$$(1)\quad \bigwedge x(y = x \leftrightarrow Ax),$$

which clearly does not express the intended meaning. And further, it is often useful to be able to symbolize propositions of the form 'the (only) A-thing is a B-thing'. This can be done on the basis of (1), but in two different ways which although equally justified, are not logically equivalent, viz:

$$(2)\quad \bigvee y(\bigwedge x(y = x \leftrightarrow Ax) \wedge By),$$

i.e. there is a thing which is (the only) A-thing and simultaneously a B-thing,

$$(3)\quad \bigwedge y(\bigwedge x(y = x \leftrightarrow Ax) \to By),$$

i.e. every thing which is (the only) A-thing is a B-thing.

On the assumption that there is exactly one A-thing both formulas express that this has the property B, but they do not have the form B ..., suggested by natural-language usage. For this reason special terms, viz. $\iota x \mathsf{A}(x)$ [to be read: 'the (only) x, *for which* A *holds for* x'] have been introduced. The variable x, which is free in $\mathsf{A}(x)$, is bound by the ι-operator [just as it is by $\bigwedge x$ or $\bigvee x$, cf. III 2, p. 54, and ι may therefore be introduced equally as a functor applicable to $x\mathsf{A}(x)$, (or to $[x \mid \mathsf{A}(x)]$, cf. V 2, p. 94)]. The expression $\iota x \mathsf{A}(x)$ is a *term* (cf. III 2, p. 53) and may be used like any other term in the construction of terms and formulas. However, the term-substitution rule TS discussed in IV 2, p. 75 must not be generally extended to 'ι-terms' (but cf. below (10), (11), (12)). Thus admitted constructions are: $\iota x Cxy$, $\iota y Cxy$, or to give an example from arithmetic, $\iota y(x+y=z)$ [usually abbreviated to '$z-x$'].

According to Russell [1] ch. 16, the general use of the ι-operator may be regulated by simply introducing $B\iota x\mathsf{A}(x)$ as abbreviation for the formula $\bigvee y(\bigwedge x(y=x \leftrightarrow \mathsf{A}(x)) \wedge By)$. However, in this case, we must not substitute $\mathsf{B}(...)$ for B ..., since if we did, it would not be clear whether $\neg B\iota xAx$ stood for

(4) $\qquad \neg \bigvee y(\bigwedge x(y = x \leftrightarrow Ax) \wedge By)$

or for

(5) $\qquad \bigvee y(\bigwedge x(y = x \leftrightarrow Ax) \wedge \neg By)$.

That it would be wrong to assume the general validity of (4) $\leftrightarrow$ (5) is shown by the following deduction: Premise: (4)$\leftrightarrow$(5). On the basis of $\neg\mathsf{A}\leftrightarrow\mathsf{B} \models \mathsf{A} \vee \mathsf{B}$, it follows that $\bigvee y(\bigwedge x(y=x\leftrightarrow Ax) \wedge By) \vee \bigvee y(\bigwedge x(y=x\leftrightarrow Ax) \wedge \neg By)$ and because of $\bigvee y\mathsf{A} \vee \bigvee y\mathsf{B} \models \bigvee y(\mathsf{A} \vee \mathsf{B})$ we have

$$\bigvee y((\bigwedge x(y = x \leftrightarrow Ax) \wedge By) \vee (\bigwedge x(y = x \leftrightarrow Ax) \wedge \neg By).$$

This gives us, by way of a propositional-logic transformation within the expression:

$$\bigvee y \bigwedge x(y = x \leftrightarrow Ax).$$

Thus our premise leads to the conclusion that an arbitrary property A holds for exactly one object, e.g. the property A defined by $\bigwedge x(Ax \leftrightarrow x \neq x)$, which is absurd.

Assuming the 'legitimacy' of ιxAx:

(6) $\qquad \bigvee y \bigwedge x(y = x \leftrightarrow Ax)$,

(4) and (5) are equivalent, i.e. (6) $\models$ (4)$\leftrightarrow$(5). The analogue for the general case can be demonstrated only step by step *via* the construction of B, cf. Whitehead–Russell [1], pp. 184–186.

For this reason the ι-operator is sometimes introduced as a special basic concept, whose use may be regulated by a suitable schema of the ι-axioms, thus e.g. the following:

(7) $\bigwedge x(y = x \leftrightarrow \mathsf{A}(x)) \wedge \mathsf{B}(y) \rightarrow \mathsf{B}(\iota x \mathsf{A}(x))$.[7]

It is fairly easy to see that this represents on the one hand a *weakening* of the equivalence

(8) $\bigvee y(\bigwedge x(y = x \leftrightarrow \mathsf{A}(x)) \wedge By) \leftrightarrow B\iota x\mathsf{A}(x)$,

which follows from Russell's definition for $B\iota x\mathsf{A}(x)$; and on the other hand a *strengthening*, since the generalizations from By to $\mathsf{B}(y)$, admissible in this direction, are already included.

The most important derivations from (7) are (9) and (10):

(9) $\bigvee y \bigwedge x(y = x \leftrightarrow \mathsf{A}(x)) \rightarrow \mathsf{A}(\iota x \mathsf{A}(x))$.

Let us illustrate this:

In the domain of integers we define $(z-x)$ by $\iota y(x+y=z)$; then by virtue of (9) it holds that $x+(z-x)=z$. It is precisely in demonstrations like this one that we require the 'trivial' statement that the only thing with a certain property *has* that property.

(10) $\bigvee y \bigwedge x(y = x \leftrightarrow \mathsf{A}(x)) \wedge \bigwedge y \mathsf{B}(y) \rightarrow \mathsf{B}(\iota \mathsf{A}(x))$.

This theorem enables us to apply a universal proposition to an object described by an individual description, since we have introduced no *general* rule for the substitution of singular descriptive terms for free variables.[8] The following rules, in particular, are derivable from (10):

(11) $\bigvee y \bigwedge x(y = x \leftrightarrow \mathsf{A}(x)) \vdash \bigwedge y \mathsf{B}(y) \rightarrow \mathsf{B}(\iota x \mathsf{A}(x))$,
(12) $\bigvee y \bigwedge x(y = x \leftrightarrow \mathsf{A}(x)) \rightarrow \mathsf{B}(z)$
$\vdash \bigvee y \bigwedge x(y = x \leftrightarrow \mathsf{A}(x)) \rightarrow \mathsf{B}(\iota x \mathsf{A}(x))$.

This last rule shows that if a requirement of 'legitimacy' is made, singular descriptive terms may be substituted for free variables like other terms.

2. DESCRIPTIONS OF ATTRIBUTES AND FUNCTIONS

A. *Attributes*

We have found it useful on a number of occasions to be able to refer to 'the property of x that is described by a condition $\mathsf{A}(x)$' or 'the relation between x and y that is described by a condition $\mathsf{C}(x, y)$', ... in general: 'the attribute whose applicability to a system $x_1, \ldots x_n$ of objects is described by a condition $\mathsf{C}(x_1, \ldots, x_n)$'.[9] This suggests an extension of predicate logic. The formula $\mathsf{C}(x, y)$ – to take the second-simplest example – is not suitable for this purpose, since for given values x, y, it already represents the relation holding between those particular values. Furthermore, there would be no possibility of distinguishing between 'the relation $\mathsf{C}(x, y)$' as such, and that property of x which, at a given value of y, is also represented by $\mathsf{C}(x, y)$ under the same convention, i.e. the property A with the 'property'[10] $\wedge x(Ax \leftrightarrow \mathsf{B}(x, y))$. Admittedly, this property A could be designated by '$\iota A \wedge x(Ax \leftrightarrow \mathsf{B}(x, y))$' on the basis of an obvious extension of V 1, p. 91. However, the derivation of the rules governing the use of such predicates would present some difficulties, as we would first have to develop a calculus of identity with formulas like $A^i_k = A^i_l$. It is simpler to extend the language of predicate logic by means of special predicates for the description of compound attributes, such as we have already used on earlier occasions. Instead of the notation $x\mathsf{A}$ or $[x]\mathsf{A}$ introduced in II 2, p. 39, in connexion with the quantifiers $\wedge x$ and $\vee x$, we shall use the predicates

$$[x \mid \mathsf{A}],\ [xy \mid \mathsf{A}],\ \ldots \text{ etc.},$$

thus adopting the symbolism most widely employed in mathematics for this purpose.[11]

(1) The required extension of $\mathbf{L}_\mathrm{F}$ may be described by the following addition to III 2, C(1)–(11), p. 53 f.:

(*) if A is a *wff*, then $[x_1 \ldots x_n \mid \mathsf{A}]$ is an n-place predicate, i.e. a *comprehensor* predicate.

As this involves extending (6) similarly, we obtain a 'simultaneous definition' of predicates and *wffs*.[12] This results in particular in an increased range of applicability of III 2, C(7)–(10), as atomic formulas (in the extended sense) such as $[ab \mid \mathsf{A}]\mathsf{t}_1\mathsf{t}_2$ can now be formed. The operator $[x_1 \ldots x_n \mid \ldots]$ is also known as abstraction operator or comprehensor because of the important part it plays in the formation of new concepts.[13] The variables $x_1, \dots x_n$ must be regarded as *bound* in the same sense as in III 2, p. 54, at all occurrences in $[x_1 \dots x_n \mid \mathsf{A}]$. In the simplest cases of (*) the variables being free in A will be precisely $x_1, \dots, x_n$. However, we must allow for the case where the course-of-value of the attribute to be described does not depend on all variables listed in the operator $[x_1 \dots x_n \mid \dots]$, as e.g. in $[xy \mid A^1x]$. On the other hand, the attribute as a whole may depend on other variables as well, as in $[x \mid B^2xy]$. Hence the general formulation.

That we must regard as bound the variables listed by $[x_1 \dots x_n \mid \dots]$, is shown also by the following interpretation of comprehensor predicates: Let $\mathfrak{B}^*([x_1 \dots x_n \mid \mathsf{A}])$ be the attribute that holds for arbitrary $\mathfrak{x}_1 \dots \mathfrak{x}_n$, if and only if[14]

$$\left(\begin{pmatrix} x_1 \dots x_n \\ \mathfrak{x}_1 \dots \mathfrak{x}_n \end{pmatrix} \mathfrak{B}\right)^*(\mathsf{A}) = \mathbf{T},$$

or strictly in symbols:

$$(2) \qquad \mathfrak{B}^*([x_1 \dots x_n \mid \mathsf{A}])(\mathfrak{x}_1, \dots, \mathfrak{x}_n) = \left(\begin{pmatrix} x_1 \dots x_n \\ \mathfrak{x}_1 \dots \mathfrak{x}_n \end{pmatrix} \mathfrak{B}\right)^*(\mathsf{A}).$$

Whatever is assigned to $x_1, \dots, x_n$ by $\mathfrak{B}$ itself, is thus of no consequence at all.

According to the definition of general validity based on this interpretation, all formulas of the form

$$(3) \qquad [x_1 \dots x_n \mid \mathsf{A}]x_1 \dots x_n \leftrightarrow \mathsf{A}$$

immediately prove to be generally valid, and this proof easily extends to their generalizations:

$$(4) \qquad [x_1 \dots x_n \mid \mathsf{A}]\mathsf{t}_1 \dots \mathsf{t}_n \leftrightarrow \mathsf{A}(x_1/\mathsf{t}_1, \dots, x_n/\mathsf{t}_n).$$

Comprehensor predicates are largely characterized by (3): if they designate anything at all so that (3) is generally valid, then this must be the attribute described by (2). It is therefore an obvious next step, when enlarging the predicate calculus for the language of predicate logic extended according to (1), to introduce as additional axioms precisely formulas (3): the calculus thus defined is complete.[15] For example, (4) is derivable from (3) by applying the rule of term-substitution TS (IV 2, p. 75).

We end this section by giving below a number of examples of concepts whose formation can be symbolically represented with the aid of comprehensors; our use of the symbolic notation will be somewhat freer.

(5) $\mathbf{n}A =_{\mathrm{Df}} [x \mid \neg Ax]$ is the property complementary to A in the sense of I 3, p. 18. According to (3) it thus holds that $\mathbf{n}Ax \leftrightarrow \neg Ax$; however, this must not be taken to mean $\mathbf{n} = \neg$: we could have written more clearly $(\mathbf{n}A)x \leftrightarrow \neg(Ax)$.

(6) $\{x_1, ..., x_n\} =_{\mathrm{Df}} [y \mid y = x_1 \vee ... \vee y = x_n]$ is the property that holds for precisely the objects $x_1, ..., x_n$ or the set consisting of exactly the objects $x_1, ..., x_n$.

(7) $A \cap B =_{\mathrm{Df}} [x \mid Ax \wedge Bx]$ is the property of having the properties A and B at the same time. For example, out of the properties of *being red* and *being a ball*, we form the property of *being a red ball.*

(8) $A \cup B =_{\mathrm{Df}} [x \mid Ax \vee Bx]$ is the property of possessing at least one of the properties A or B. Thus out of the properties of being a *son* or a *daughter* (of a specific parental pair b) we form the property of being a *child* (of b).

(9) $[xy \mid Ryx]$ is the relation that holds between x and y if and only if R holds between y and x. Thus out of the relation of *being the superior of* ... we form the relation of *being the subordinate of* ...

(10) $[xy \mid \bigvee z(Rxz \wedge Szy)]$ is the relation that holds between x and y if and only if there is a z with Rxz and Szy. For example, if R is the relation $[xy \mid x$ *child of* $y]$, then $[xy \mid \bigvee z(Rxz \wedge Rzy)]$ is the relation $[xy \mid x$ *grandchild of* $y]$. Other family relationships may be expressed similarly, if need be by longer 'concatenations'.

(11) $[xy \mid y = fx]$ is the formal description of the representation in graph-form (briefly: of the graph) of the function f in the 'x, y plane'.[16] The reader should call to mind graphs of simple functions such as $[xy \mid y = 2x + 3]$, $[xy \mid y = x^2]$.)

With the aid of the comprehensors so far introduced we may also describe the combination of two functions f and g. However, it must be borne in mind that in employing this procedure we form the *graph* of the compound function from the given *functions*, e.g.:

(12) $[xy \mid y = fx + gx]$, $[xy \mid y = fx \cdot gx]$, $[xy \mid y = fgx]$.[17]

B. *Functions*

Within the framework of set theory functions are often identified with their graphs – i.e. we say that a function is nothing other than a pair-set (etc.) with specific properties. In fact, however, the concept of a function has the same generality as that of a set; in other words, sets may be introduced as functions with specific properties, viz as attributes. This being the case, however, it becomes somewhat artificial to describe functions by means of graphs. It seems more appropriate to extend the use of comprehensors by introducing 'comprehensor functors' in addition to comprehensor predicates; i.e. comprehensor terms that designate functions directly (instead of their graphs). And just as in the case of comprehensor predicates we write a formula (for a truth value) after the comprehensor, we now write a term (for the value of the function) after the comprehensor.

On the basis of II 2, p. 39, we should obtain comprehensor functors like $[x]\mathtt{t}$, $[x_1 \ldots x_n]\mathtt{t}$. We prefer, however, to use a symbolism analogous to (1), p. 94, and therefore make the following additions to III 2, C (p. 53 f.):

(1) If $\mathtt{t}$ is a term, then $[x_1 \ldots x_i \mid \mathtt{t}]$ is an i-place *functor;*
(2) If Φ is an i-place functor, then $\Phi\mathtt{t}_1 \ldots \mathtt{t}_i$ is a term.[18]

We can now construct terms, or object names, having the form $[x_1 \ldots x_i \mid \mathtt{t}]\mathtt{t}_1 \ldots \mathtt{t}_i$ and having the same meaning as $\mathtt{t}(x_1/\mathtt{t}_1, \ldots, x_i/\mathtt{t}_i)$ – i.e. as $\mathtt{t}$, where the variables $x_1, \ldots, x_i$ are simultaneously replaced by the terms $\mathtt{t}_1, \ldots, \mathtt{t}_i$. This requires the following supplement to the definition of $\mathfrak{B}^*$ given in III 3, C, p. 59 f.:

(3) $$\mathfrak{B}^*([x_1 \ldots x_i \mid \mathtt{t}])(\mathfrak{x}_1, \ldots, \mathfrak{x}_i) = \left(\begin{pmatrix} x_1 \ldots x_i \\ \mathfrak{x}_1 \ldots \mathfrak{x}_i \end{pmatrix}\mathfrak{B}\right)^*(\mathtt{t}).$$

Accordingly all equations of the form

(4) $$[x_1 \ldots x_i \mid \mathtt{t}]x_1 \ldots x_i = \mathtt{t}$$

are generally valid. The attaching of variables to the functor – to express the application of the function to its arguments – in a sense reverses the process of abstraction or comprehension.

Analogously to the extension of the predicate calculus by means of the axioms (3), p. 95, we obtain a complete calculus for the language $\mathbf{L}_{\mathrm{I}}$ extended by B, (1) and (2), if we supplement the axioms of IC by B, (4). For example, by substitution in (4) we immediately obtain the schema that describes the 'application' of a comprehensor functor to arbitrary arguments:[19]

(5) $$[x_1 \ldots x_n \mid \mathfrak{t}]\mathfrak{t}_1 \ldots \mathfrak{t}_n = \mathfrak{t}(x_1/\mathfrak{t}_1, \ldots, x_n/\mathfrak{t}_n).$$

To illustrate the use of comprehensor functors we give below the definitions of the compound functions whose graphs are described in A (12). We define:

(6) $$f + g = [x \mid fx + gx],^{20} \quad f \cdot g = [x \mid fx \cdot gx],$$
$$f \circ g = [x \mid fgx].$$

The usual definitions of $f+g, f \cdot g, f \circ g$ are:

(7) $$(f + g)x = fx + gx, \quad (f \cdot g)x = fx \cdot gx, \quad (f \circ g)x = fgx.$$

On the other hand, in (1), (2) and (4), we have incorporated the general method contained in (7).

Another important application of this method is forming the *converse function* of a function. Here we additionally require the singular description terms introduced in V 1, B, p. 91. Converse functions can, of course, only be formed out of reversible functions f, i.e. f must satisfy the condition $\bigwedge x_1 \bigwedge x_2 (fx_1 = fx_2 \rightarrow x_1 = x_2)$. For those y that satisfy the condition $\bigvee x(y = fx)$, there is exactly one x so that $y = fx$, and this is usually designated as $f^{-1}y$.[21]

The 'converse function of f' thus introduced may be described as follows with the methods so far developed. In the first place $f^{-1}y$ is that x for which $y = fx$, i.e. (according to V 1, B, p. 93) $\iota x(y = fx)$. Since according to V 1 this expression may be used as a term on the condition that $\bigvee z \bigwedge x(z = x \leftrightarrow y = fx)$, we now form the comprehensor functor $[y \mid \iota x(y = fx)]$ and write f^{-1} as abbreviation for it.

Taking into account V 1, B (12), p. 93, we then obtain the following theorem[22] as a counterpart to (4):

(8) $\bigvee z \bigwedge x(z = x \leftrightarrow y = fx) \rightarrow f^{-1}y = \iota x(y = fx)$.

For those y for which it holds that they occur as values of f, e.g. for $y=fu$, the premise in (8) may be weakened, and we obtain:

(9) $\bigvee z \bigwedge x(fu = fx \rightarrow z = x) \rightarrow f^{-1}fu = \iota x(fu = fx)$.

Keeping the premise, we obtain also the following implications:

(10) $\bigvee z \bigwedge x(fu = fx \rightarrow z = x) \rightarrow \iota x(fu = fx) = u$,

whence the simpler formula

(11) $\bigvee z \bigwedge x(fu = fx \rightarrow z = x) \rightarrow f^{-1}fu = u$,[23]

whose premise is a consequence of the general reversibility of f, since it expresses precisely the reversibility 'at fu'.

The above-mentioned applications of abstraction or comprehension operators are, of course, merely examples of the way in which the introduction of these operators allows us explicitly to designate abstract objects which otherwise could only be described in terms of their properties.

3. MANY-SORTED THEORIES. CONCEPTS AS OBJECTS

In applications of logic we rarely find ourselves dealing with the expressions 'for all things (whatsoever) ...' or 'for at least one thing ...', but rather with phrases like 'for all animals ...', 'for all points ...', etc. Such turns of phrase can be expressed within the language of predicate logic, as we have seen in II 2, p. 39, but for a number of purposes it is advisable to bring the symbolic notation closer to natural-language usage.

This is done by introducing for each of the types of things in question – such as *points*, *straight lines* and *planes* or in general $\mathbf{D}_1, \mathbf{D}_2, \ldots, \mathbf{D}_i$ – special kinds of variables – such as $p_1, p_2, \ldots, g_1, g_2, \ldots, e_1, e_2, \ldots$, or in general e.g. $a_{i1}, a_{i2}, \ldots, a_{ik}$, for the objects in $\mathbf{D}_i$. Out of these the formulas of a 'many-sorted' language are formed exactly as described in III 2, C, p. 52 f. for the general logic of predicates. These *wffs* deviate from the general predicate logic interpretation only in that we read:

(1) $\bigwedge a_{ij}\mathsf{A}$ as: for all objects in $\mathbf{D}_i$ (it holds that) A;
(2) $\bigvee a_{ij}\mathsf{A}$ as: for at least one object in $\mathbf{D}_i$ (it holds that) A.

Sometimes the predicate variables (and, where provided, the function variables) are also specified in the sense that the sort of object-variable and, where applicable, the sort of terms that may occur at each position, are fixed. For example, if A_b^{rst} stands for a three-place predicate variable whose argument places are reserved for the variables a_{ri}, a_{sj}, a_{tk} (in this sequence), then we may construct out of them the atomic formulas $A_b^{rst}a_{ri}a_{sj}a_{tk}$.

Taking (1) and (2) into account the semantic concepts of general validity and derivability may be transferred in an analogous sense from III 3, p. 59 f. to the language of a many-sorted theory, and similarly we obtain a many-sorted calculus by analogous transference of the axioms and rules of the predicate calculus.

To give the reader an example of the resultant increased legibility of formulas, we write out below – in a suitable many-sorted language – the geometric axiom formulated in III 2, p. 56 in the language of general predicate logic:

(3) $\qquad \bigwedge p_1 \bigwedge p_2 \bigvee g(\mathbf{L}p_1 g \wedge \mathbf{L}p_2 g).$

Whilst our immediate aim in introducing the idea of many-sorted theories has been to give more convenient expression to propositions already formulable in $\mathbf{L}_F$, it can also be used to extend considerably the language of predicate logic. Such extension is required if we wish to give a systematic symbolic expression to the conception of attributes (or sets) and functions as objects – a conception that has already proved unavoidable on a number of occasions above when we wished merely to be able to talk about them.[24]

The introduction of comprehensor terms (cf. V 2, p. 94 and 97) does not in itself bring about this extension. The first decisive step in this direction is rather the introduction of 'For all ...' and 'There is (are) ...' in respect of attributes or sets (and, where applicable, functions), e.g. of the formulas $\bigwedge A_k^i \mathsf{A}$ and $\bigvee A_k^i \mathsf{A}$. These formulas can be interpreted like the formulas of a many-sorted theory, or logic, where apart from the sort $\mathbf{D}_*$ of objects in the narrower sense, we have the following:

the sort $\mathbf{D}_0$ of truth values, with the variables A_k^0,

the sort $\mathbf{D}_1$ of one-place attributes over $\mathbf{D}_*$, with the variables A_k^1, and *in general:*

the sort $\mathbf{D}_i$ of i-place attributes over $\mathbf{D}_*$, with the variables A_k^i.

Instead of predicate variables for attributes *over* the domains $\mathbf{D}_*$, $\mathbf{D}_0$, $\mathbf{D}_1$, $\mathbf{D}_2$, ..., $\mathbf{D}_i$, ... we here have only some specific predicates determined by the 'nature' of the above sorts. The fact that an object from $\mathbf{D}_1$ is *applicable* to an object from $\mathbf{D}_*$, gives rise to a natural *two*-term relation between the $\mathbf{D}_1$-objects and the $\mathbf{D}_*$-objects; the fact that an object from $\mathbf{D}_2$ is applicable to a pair of objects from $\mathbf{D}_*$, gives rise to a natural *three*-term relation between $\mathbf{D}_2$-objects, $\mathbf{D}_*$-objects and $\mathbf{D}_*$-objects; etc. We could introduce corresponding predicates for these special attributes, e.g. ϕ^2, ϕ^3, ..., with the atomic formulas $\phi^2 A_k^1 a_j$, $\phi^3 A_k^2 a_{j_1} a_{j_2}$, or in general $\phi^{i+1} A_k^i a_{j_i} \dots a_{j_i}$. If as is usual, we omit the (predicates) ϕ^i but still refer to their interpretations (as if they were *invisible* predicates) we have the same atomic formulas as in $\mathbf{L}_\mathrm{F}$. This is a way of describing the intention that what is indicated by a variable A_k^i is an object as well as a function (in the former version A_k^i gave the objective, and ϕ^{i+1} the functional part of it). The set-theory notation $a_j \in A_k^1$, $(a_{j_1} a_{j_2}) \in A_k^2$, ... is also used occasionally, and so is the form $A_k^1 \ni a_j$, $A_k^2 \ni (a_{j_1} a_{j_2})$, ..., which represents a compromise between the symbolism of predicate logic and that of set theory.[25]

By means of these or similar conventions, the essential feature of which is always the introduction of $\wedge A_k^i$ and $\vee A_k^i$ (in some cases also of $\wedge f_k^i$ and $\vee f_k^i$), a second-order language $\mathbf{L}_\mathrm{F}^2$ of predicate logic is formed.

The minimum requirement for the interpretation of the formulas of $\mathbf{L}_\mathrm{F}^2$ is that for given domains $\mathbf{D}_*$, $\mathbf{D}_0$, $\mathbf{D}_1$, $\mathbf{D}_2$, ... the predicates ϕ^2, ϕ^3, ... (or their analogues) have fixed meanings. The requirement is normally met if all domains $\mathbf{D}_i$ consist of attributes (specifically, of *i*-place attributes) over $\mathbf{D}_*$. This in itself distinguishes the second-order predicate logic from a general many-sorted logic and makes it a part of a logic (or theory) of types, where 'type' refers to a hierarchy of abstract 'objects' which are the outcome of our iterated objectification of functions. Further, general validity and other semantic conepts are usually defined in such a way as to coincide, for formulas already belonging to $\mathbf{L}_\mathrm{F}$, with the concepts defined for $\mathbf{L}_\mathrm{F}$. This is always the case if the requirement is made that $\mathbf{D}_i$ consist of *all* *i*-place attributes over $\mathbf{D}^*$. These interpretations have been designated as 'absolute', 'standard', 'normal' or 'maximal'.

It is unlikely that we would consider other interpretations of $\mathbf{L}_\mathrm{F}^2$, had it not been shown that for infinite $\mathbf{D}_*$ each axiomatic description of the appurtenant (maximal) $\mathbf{D}_i$ as maximal is insufficient. It can be demonstrated that even a catalogue of all propositions formulable in $\mathbf{L}_F^2$ that are 'true for $\mathbf{D}_*$, $\mathbf{D}_0$, $\mathbf{D}_1$, $\mathbf{D}_2$, ...' would be inadequate for this purpose. For according to a theorem of Löwenheim and Skolem such a catalogue – which might be regarded as a kind of super axiomatic system – would always have a model with the same $\mathbf{D}_*$ (and $\mathbf{D}_0$) but non-maximal domains $\mathbf{D}_1, \mathbf{D}_2, \ldots$. Apart from this, it is not possible to describe such a catalogue by means of a calculus in the sense of IV 2, p. 70 f. (as will be shown in VII 3, p. 135), so that an 'axiomatic' description of the domains $\mathbf{D}_1$, $\mathbf{D}_2$, ... presents difficulties on two scores.

Some researchers have gone so far as to advocate that these conceptions be excluded as senseless from the field of logic and mathematics. Nevertheless, the general practice is to interpret $\mathbf{L}_F^2$ formulas as 'normal', since their meaning might otherwise be subject to imponderable changes.

For example, from the possible definition of identity in $\mathbf{L}_F^2$, viz:

$$(4) \qquad x = y =_{\mathrm{Df}} \bigwedge A^1(A^1x \rightarrow A^1y)$$

we can demonstrate, in a calculus appropriate to $\mathbf{L}_F^2$, the axioms of identity formulated for $\mathbf{L}_I$ in V 1, A, p. 88. However, this does not exclude the possibility that in a non-normal interpretation compatible with the axioms, the relation designated by '=' holds between two different objects $\mathfrak{x}$, $\mathfrak{y}$ out of $\mathbf{D}_*$, e.g. if all the properties in respect of which $\mathfrak{x}$ and $\mathfrak{y}$ differ are absent from $\mathbf{D}_1$. And this is a comparatively innocuous example. Further, the definition contained in (4) shows that $\mathbf{L}_F^2$ may be based equally on $\mathbf{L}_F$ or on $\mathbf{L}_I$.

It is an obvious next step to extend $\mathbf{L}_F^2$ for all of its variables, i.e. for all provided types by adding the expressions described in V 1 for $\mathbf{L}_F$. This also gives us the possibilities of expression contained in V 2, p. 94, as is shown by the following definitions:

$$(5.1) \qquad [x \mid \mathsf{A}(x)] =_{\mathrm{Df}} \iota A^1 \bigwedge x(A^1x \leftrightarrow \mathsf{A}(x)),$$
$$(5.2) \qquad [xy \mid \mathsf{B}(x, y)] =_{\mathrm{Df}} \iota A^2 \bigwedge xy(A^2xy \leftrightarrow \mathsf{B}(x, y)), \text{ etc.}$$

We give below a typical example of what can be said with $\mathbf{L}_F^2$, using the extensions introduced in V 1 and 2, for the sake of simplicity.

The generalization of

$$(6) \qquad [xy \mid Axy], [xy \mid Axy \vee \bigvee z(Axz \wedge Azy)],$$
$$[xy \mid Axy \vee \bigvee z(Axz \wedge Azy) \vee \bigvee z \bigvee u(Axz \wedge Azu \wedge Auy)],$$

which is only incompletely expressed by means of '...' in

(7) $[xy \mid Axy \vee \bigvee z(Axz \wedge Azy) \vee \bigvee z \bigvee u(Axz \wedge Azu \wedge Auy) \vee \ldots]$[26]

may be written in 'closed' form in $\mathbf{L}_\mathrm{F}^2$ as follows:

(8) $[xy \mid \bigwedge B(\bigwedge uv(Auv \rightarrow Buv) \wedge \bigwedge uvw(Buv \wedge Avw \rightarrow Buw) \rightarrow Bxy)]$.

Whereas only finite 'A-chains' of a determinate length can be described in terms of (6), (8) yields the definition of 'A-chains of arbitrary finite length' and therewith the definition of finiteness which is generally used for the definition of natural numbers in terms of logic. The concept of finiteness provides a vicious example of how concepts can be twisted if $\mathbf{L}_\mathrm{F}^2$ is used in the sense of a non-normal interpretation: under each intuitively correct definition of finiteness, some infinite set might pass as finite.

In our description of $\mathbf{L}_\mathrm{F}^2$ we have so far not touched on the possibility of talking about arbitrary properties of (and arbitrary relations between) attributes, i.e. about attributes over the domains $\mathbf{D}_1$, $\mathbf{D}_2$, ... These 'second-order attributes' (from the point of view of ordinary predicate logic) are the first-order attributes of the special many-sorted theory, as the language of which $\mathbf{L}_\mathrm{F}^2$ was at first understood. All that is required, then, is to extend $\mathbf{L}_\mathrm{F}^2$ by means of the predicate variables omitted above. From a formal point of view this gives us a many-sorted language as introduced initially, supplemented by the special predicates ϕ^2, ϕ^3, ... (cf. p. 101).

This language may be extended as required in the same way as described above for $\mathbf{L}_\mathrm{F}$, and this process may be repeated as often as required. The need for such extensions can be demonstrated by means of examples, but we shall defer doing so until we have the appropriate symbolism at our disposal.

Clearly, the usefulness of the above extensions $\mathbf{L}_\mathrm{F}^2$, ... of $\mathbf{L}_\mathrm{F}$ depends to a great extent on our being able to order distinctly the terrifying multiplicity of new types created. The logic that results from such repeated extensions is known as higher-order logic or theory of types.

The various types are designated by type indices (often referred to simply as types), and the rules of formation for the type indices yield the required principle of order. Thus if τ is a type, then $\mathbf{D}_\tau$ is the domain of objects 'of type τ', and a_1^τ, a_2^τ, ... are variables[27] 'of type τ'. That is, for all value assignments $\mathfrak{B}$, to be introduced as in III 3, p. 59, $\mathfrak{B}(a_i^\tau)$ belongs to $\mathbf{D}_\tau$.

Let $\mathbf{D}_0$ again be the domain of truth values and $\mathbf{D}_*$ the domain of given objects. (These might be actual objects or the outcome of a previous abstraction.) The 'higher' types must be formed from the types o and $*$, and various ways of doing this have been suggested. The following procedure (of *Principia Mathematica*) is adequate for the above-mentioned extension of $\mathbf{L}_\mathrm{F}$, with the exclusion of all function-types other than attributes. We define a language $\mathbf{L}_{(9)}$:

(9) If $\alpha_1 \ldots \alpha_n$ are types, then let $(\alpha_1 \ldots \alpha_n)$ be the type of those attributes whose first position variable refers to objects in $\mathbf{D}_{\alpha_1}$ and ... and whose nth-position variable refers to objects in $\mathbf{D}_{\alpha_n}$.

We thus designate the types already occurring in $\mathbf{L}_\mathrm{F}^2$ as follows:

(10) $*, \mathrm{o}, (*), (**), (***), \ldots$ [28]

And we can also form types such as, for example:

(11) $(\mathrm{o}*), (*(*)), (**((*)))$

or, somewhat more systematically:

(12) $*, (*), ((*)), (((*))), \ldots.$

This wealth of possibilities is less confusing than might appear at first sight, since we can select from it whatever is required by any one particular problem. It would, however, be entirely arbitrary to stop the proliferation of types at any one point. Again, it is relatively simple to formulate general principles, since account need be taken only of the formation procedure (9), and not of the host of possibilities. Thus the decisive step in the formation of the language $\mathbf{L}_{(9)}$ of the theory of types *determined by* (9), is the formation of the atomic formulas described by [29]

(13) For arbitrary types $\alpha_1 \ldots \alpha_n$, the sequence built out of variables $a_k^{(\alpha_1 \ldots \alpha_n)}\, a_{i_1}^{\alpha_1} \ldots a_{i_n}^{\alpha_n}$ constitutes a formula. In all other respects we proceed as for $\mathbf{L}_\mathrm{F}$ (with the exclusion of the function variables).

As Kuratowski has suggested, everything that can be said in $\mathbf{L}_{(9)}$ can also be said in a language $\mathbf{L}_{(12)}$ restricted to types (12). The language $\mathbf{L}_{(12)}$ is much simpler to describe, but on the other hand the required definitions are correspondingly longer. Nevertheless, it is sometimes useful to have a language which has so narrow a range of vocabulary and is yet so expressive (i.e. one in which only *one*-place predicate variables occur).

Using the method indicated in V 2, (11), p. 96, it is possible to describe functions of all kinds in terms of $\mathbf{L}_{(9)}$ or $\mathbf{L}_{(12)}$. It is, however, advisable to treat the different kinds of functions other than attributes as separate types.

This means that convention (9) would have to be modified, for we would have to symbolize the type of the function value which we had been able to suppress in (9), since it was always 'o'.

We thus stipulate:

(14) If $\alpha_0, \alpha_1, \ldots, \alpha_n$ are types, then $(\alpha_0 \mid \alpha_1 \ldots \alpha_n)$ is the type of the functions with the definition domains $\mathbf{D}_{\alpha_1}, \ldots \mathbf{D}_{\alpha_n}$ and values in $\mathbf{D}_{\alpha_0}$.
In this way all earlier types $(\alpha_1 \ldots \alpha_n)$ are preserved in the form $(o \mid \alpha_1 \ldots \alpha_n)$; but other types are added, in particular types of functions whose arguments and values are again functions – which is becoming increasingly important in modern mathematics.

As Schönfinkel has shown, these new types can also be used to reduce all types to the types of one-place functions, and this in a manner much simpler than that of Kuratowski above. For example, the type $(((\alpha \mid \beta) \mid \gamma) \mid \delta)$ clearly expresses the same as the type $(\alpha \mid \delta\gamma\beta)$, as is shown by the corresponding application of (13). On the one hand we form

(15) $\qquad a_1^{(\alpha \mid \delta\gamma\beta)} a_2^{\delta} a_3^{\gamma} a_4^{\beta},$

and on the other, step by step[30]

(16) $\qquad a_1^{(((\alpha \mid \beta) \mid \gamma) \mid \delta)}, a_2^{\delta}, a_3^{\gamma}$ and $a_4^{\beta},$

whereupon the result is normalized. The general formulation follows fairly obviously.

We now stipulate in place of (14):

(17) If α, β are types, then $(\alpha\beta)$ is the type of the functions with arguments in $\mathbf{D}_{\beta}$ and values in $\mathbf{D}_{\alpha}$.[31]

We may then describe the language $\mathbf{L}_{\mathrm{T}}$ of the theory of types in terms of the following rules:

(18) Every constant of type α is an expression of type α (this schema

requires specification of the constants that are to be effectively introduced);

(19) Every variable of type α is an expression of type α;

(20) For arbitrary expressions $A^{(\alpha\beta)}$, B^{β} of types $(\alpha\beta)$ or β, $(A^{(\alpha\beta)}B^{\beta})$ is an expression of type α.

Expressions of type o will be counted as *formulas* in the earlier sense. All other expressions will be regarded as *terms* in the extended sense.

Within the framework of (17)–(20) we can further introduce all possible logical functions as constants of corresponding types. An extremely elegant calculus for a language of this kind has been developed by Alonzo Church [2], its usefulness deriving largely from the fact that as well as (18)–(20), the comprehensor terms of all possible types are admitted in virtue of the following addition to (19) and (20):

(21) If A^{α} is an expression of type α, then $[a_i^{\beta} \mid A^{\alpha}]$ is an expression of type $(\alpha\beta)$.[32]

The language of the theory of types enables us, in particular, to establish the connexion between the use of numbers for counting objects and the abstract use of natural numbers in calculating.

In the simplest case numbers may be regarded as properties of objects of type $(o*)$,[33] i.e. *of properties* of objects[34] and not of objects themselves (imagine a 'three-Magi' or a 'seven-dwarf'). They are thus objects of type $(o(o*))$. Now these objects out of $\mathbf{D}_{(o(o*))}$ may be described in purely logical terms, for we can express without using numbers that two objects out of $\mathbf{D}_{(o*)}$ hold for the same number of $\mathbf{D}^*$-objects. It is further possible to define *addition* and *multiplication* for numbers as specific objects in $\mathbf{D}_{(((o(*))(o(o*)))(o(o*)))}$[35] and to prove the known laws of arithmetic on the assumption that there are sufficiently many things in $\mathbf{D}_*$.

Turning now to botany, we conclude with an example that is closer to life. Let $\mathbf{D}_*$ be the domain of all botanical individuals. Among properties over $\mathbf{D}_*$, i.e. objects out of $\mathbf{D}_{(o*)}$, would then be included all the concepts under which botanists are accustomed to order their wealth of classificatory possibilities, i.e. (from top to bottom): *divisions*, *classes*, *orders*, *families*, *genera* and *species*. The concepts *division*, *class*, *order*, *family*, *genus* and *species* would then be regarded as properties of objects out of $\mathbf{D}_{(o*)}$ or as objects in $\mathbf{D}_{(o(o*))}$.

This, however, is not how botanists actually use these concepts. Even if a genus contains only one species, or if a family has only one genus, these are distinguished. We ought therefore to regard species as occurring in $\mathbf{D}_{(o*)}$, genera in $\mathbf{D}_{(o(o*)),\ldots,}$ divisions in $\mathbf{D}_{(o(o(o(o(o(o*))))))}$. Fortunately this is a topic on which we need exercise our minds only rarely, since for every construct in $\mathbf{D}_{(o(o*))}$ there is a natural counterpart in $\mathbf{D}_{(o*)}$.

NOTES

1 In this connexion cf. V 3, p. 106 f.
2 Cf. II 2, p. 40.
3 Cf. I 3, p. 25.
4 It is generally agreed that a 'literal' translation into natural language is not very helpful.
5 Using the more abstract concepts of V 2, p. 96, we obtain e.g. $\bigvee xyz(x \neq y \wedge x \neq z \wedge y \neq z \wedge A \cap B \equiv \{x, y, z\})$, as equivalent to – and more intelligible than – the last-mentioned formula above.
6 The corresponding way of expressing that there are at least three A-things is less intuitive, and we merely state it: $\bigwedge xy \bigvee z(z \neq x \wedge z \neq y \wedge Az)$.
7 The substitution of $\iota x\mathsf{A}(x)$ for y requires precautionary measures similar to those formulated in IV 2, p. 75 for the substitution of terms.
8 Such a rule, though occasionally chosen as a basic rule, is incompatible with the maxim to which we have here adhered, viz that at the most derivations from (8) are to be demonstrable.
9 For further examples, cf. p. 96.
10 On properties of properties, etc., cf. V 3, p. 103.
11 Admittedly the mathematical use is in connexion with sets, but the two uses are very close, sets and attributes being occasionally even 'identified', as they can replace each other in appropriate formulations. The proposed use of [ ], instead of { }, is to point out that the denoted attributes are, in general, not in the universe of discourse.
12 If additionally we formalize the use of the definite article, as under V 1, p. 91, a simultaneous definition of *terms*, *predicates* and *wffs* is required.
13 From the Latin *comprehendere*. A particularly elegant comprehensor theory and technique will be found in Curry's 'Combinatory Logic' (cf. Curry–Feys [1] and Cogan [1]).
14 $\binom{x_1 \ldots x_n}{\mathfrak{x}_1 \ldots \mathfrak{x}_n}\mathfrak{B}$ is an obvious generalization of $\binom{x}{\mathfrak{x}}\mathfrak{B}$ in III 3, p. 60.
15 It is only if one proceeds to analyse $\bigwedge x\mathsf{A}$ and $\bigvee x\mathsf{A}$ into $\bigwedge[x \mid \mathsf{A}]$ and $\bigvee[x \mid \mathsf{A}]$ respectively, on the basis of II 2, p. 39 (an obvious step after the introduction of comprehensors) that small additions become necessary.
16 Whether we write '$[xy \mid y = fx]$' or '$[yx \mid y = fx]$' is of no consequence, so long as we are consistent.
17 The meaning here of 'fgx' is: f applied to gx; cf. III 2, C (2), p. 53.
18 In the sense of a generalization of III 2, C (2).
19 Cf. (4) p. 95.
20 Analogously, where appropriate, also $f + g = [xy \mid fxy + gxy]$ etc.
21 The exponent $^{-1}$ must not, of course, be confused with a place index.
22 In a semantic sense only. Calculi for a predicate logic extended by comprehensor functors $[x_1 \ldots x_n \mid \mathsf{t}]$ have up to now been little investigated. One might consider $\bigvee y(y = \mathsf{t}) \rightarrow [x_1 \ldots x_n \mid \mathsf{t}]x_1 \ldots x_n = \mathsf{t}$ (y not in t) as a possible axiom schema that would yield (4) as well as the schema indicated by (8). The premise would then be

demonstrable for 'well-behaved' terms, and would follow for ι-terms from the 'condition of legitimacy'.

[23] A counterpart that may come to the reader's mind and that holds for those u with the property $[u \mid \bigvee x(u = fx)]$, is ... $\rightarrow ff^{-1}\, u = u$.

[24] It seems likely that by introducing truth values in II 2, p. 33 f. we have created the wrong conditions for the task of systematically treating propositions and events as objects – and yet this task seems in many respects to be equally justified and even, in the light of a number of examples, advisable. Cf. also III 3 A, p. 56.

[25] Safeguards against the risk of confusion must, of course, be built into the language whenever we use the convention by which Φ^2, Φ^3, ... are replaced by the same (or an invisible) predicate.

[26] Let A be a two-place predicate variable.

[27] The type index is often written underneath; we prefer, however, to write it above, as this seems more consistent with our earlier usage.

[28] In the case where truth-values are regarded as null-place attributes, it would be appropriate to write '()' instead of 'o'.

[29] We here use the 'invisible predicates' mentioned on p. 101 as well as their generalizations.

[30] The reader has already encountered this idea in the method by which the formation of $A^n\mathbf{t}_1 \ldots \mathbf{t}_n$ is broken up into a stepwise addition of *one* term at a time; cf. III 2, C (5) and (7), p. 53.

[31] We have written the simplified form '(αβ)' instead of '(α | β)' because:

1. there is little risk of confusion with (9);
2. under the convention of writing values in the first position, functions of type (αβ) have graphs of type (αβ) in the sense of (9).

[32] Church's symbol has here been altered in the sense of V 2, p. 94.

[33] That they *need* not be thus regarded is shown precisely by Church [2] who treats them as objects of type ((∗∗)(∗∗)), or in general: of type ((αα)(αα)).

[34] Cf. the various ways discussed in V 1, p. 89 f., of expressing that one A^1 holds for exactly n things.

[35] This looks complicated but proves easier to read if we write 'n' for (o(o∗)) to obtain **D**((*nn*)*n*). We have given this example merely to show that fairly abstruse types occur even in the elementary stages of mathematics. Mathematicians will have little difficulty in finding further, quite different examples, but they too will no doubt be glad that there is no need to be constantly thinking about them.

CHAPTER VI

ANTINOMIES

Under the catch-phrase 'antinomies' we conveniently group together a whole range of problems whose common feature is that they give rise to contradictory conclusions from plausible conceptions and premises. Depending on our temperamental make-up we tend to explain the contradictions as arising from the initial conceptions, from their formulation in the premises, or from the logic employed in the deduction.

We give below a number of typical examples to show how the contradictory conclusions may be avoided by a more precise analysis of the initial conceptions and premises. In my opinion these antinomies can be resolved by this method, and any other analogously formulated antinomies will be similarly resolvable.

It seems to me that the antinomies are important because they have forced us to analyse our thought processes more clearly and to work out a more appropriate formulation of the premises, rather than because certain constructs which are important for and characteristic of modern mathematics have had to be rejected. At any rate, as will be shown below, certain conceptions have had to be corrected and, understandably, opinions diverge as to the extent to which these corrections should be generalized to guard against further mistakes.

A distinction is usually made between so-called logical antinomies, where a contradiction is formally deduced from plausible assumptions formulated in a language **L**; and semantic antinomies, where assumptions about the relation holding between the language **L** and what is expressed in it give rise to paradoxical conclusions which are then shown to be formal contradictions in the metalanguage. However, I do not consider the distinction to be very important: surely, when we judge the assumptions to be plausible in the case of the 'logical' antinomies, we also interpret the language in which the antinomy is formulated – unless we treat the whole matter as a game.

1. THE SET OF ALL THINGS WITH A GIVEN PROPERTY

It has been shown in V 3, p. 100 f. how attributes, i.e. value distributions of concepts, or their corresponding sets, i.e. the extensions of concepts, can be treated as things, viz as 'objects of our intuition or of our thought' (in the present context: as 'objects of our thought'). Instead of dividing things (in the extended sense) into types, as outlined there, it is probably simpler to talk about them as 'members' of *one* domain, which we then use as the domain of objects, the universe of discourse, for the interpretation of a suitable language of predicate logic.

Apart from the trivial null- and all-attributes, only identity is characterized as a 'natural' attribute in general predicate logic. We shall now, however, add a number of other attributes, and by introducing new constants as names for them we shall adapt the language of predicate logic to the new interpretation. Thus:

(1) The one-place attribute that holds for exactly those things that are not sets, the basic objects, will be designated by 'B';

(2) The two-place attribute that holds for a pair (x, y) if and only if x is an element ('member') of y, will be designated by 'E'. (Thus 'Exy' stands for 'x is an element of y').[1]

The domain of objects **D** in which this language can be interpreted will have to consist on the one hand of basic objects[2] and on the other, of all sets 'that can be constructed'. Several definitions of sets are possible, depending on what we mean by the verb 'construct'. *Cantor's* definition was that sets are constituted by (literally: 'By a set we understand every') *collection into a whole of definite, distinct objects of our intuition or of our thought.*[3] Starting from this definition, *Cantor* worked out a large part of what is today known as naive set theory. The antinomies which it was later found to contain, made it clear that the concept of a set had to be used with greater caution. The following is intended as a contribution towards such clarification.

According to *Cantor's* definition there is a natural connexion between the *properties* that are meaningful for the things of a domain and the *sets* of the things that have these properties. A problem only arises with the *assumption* or requirement that all these sets themselves should belong to the domain of objects in the wider sense. Let us suppose that we have succeeded in finding such a domain **D**. Then **D** will include:[4]

(0) objects in the narrower sense, also called *basic elements* in this context, i.e. things of type $*$ in the sense of V 3, p. 104;

(1.1) as an *object* in **D** the *set* b_1 of the basic elements;

(1.2) every *sub-set* of b_1, i.e. all objects of type $(o*)$ in the sense of V 3, p. 104.[5]

D now contains all objects of type $(o*)$ in the sense of V 3, p. 104. But this is not all, of course. We must also let **D** include as *objects:*

(2.1) the *set* b_2, consisting of all objects described so far;

(2.2) all *sub-sets* of b_2.

This procedure may be repeated at will. In the general case n, therefore, **D** will include

(n.1) as *object* the *set* b_n consisting of all sets previously constructed;

(n.2) as *objects* all *sub-sets* of b_n.

All sets thus occurring as objects in **D**, may also be described in terms of properties A formulated for any object in **D** whatsoever.[6]

Thus:

(1.1′) $\bigwedge x(\mathsf{E}x\mathsf{b}_1 \leftrightarrow \mathsf{B}x)$ describes b_1;

(1.2′) $\bigwedge x(\mathsf{E}x\mathsf{t}_{1A} \leftrightarrow \mathsf{E}x\mathsf{b}_1 \wedge Ax)$ 'describes' the sub-sets t_{1A} of b_1 with the aid of arbitrary attributes A [properly $\mathfrak{V}(A)$] *over* **D**:

(2.1′) $\bigwedge x(\mathsf{E}x\mathsf{b}_2 \leftrightarrow \mathsf{E}x\mathsf{b}_1 \vee x \subseteq \mathsf{b}_1)$,
[where $a \subseteq \mathsf{b}$ stands as abbreviation for $\bigwedge y(\mathsf{E}ya \rightarrow \mathsf{E}yb)$] expresses that b_2 consists of the elements and sub-sets of b_1;

(2.2′) $\bigwedge x(\mathsf{E}x\mathsf{t}_{2A} \leftrightarrow \mathsf{E}x\mathsf{b}_2 \wedge Ax)$
describes the sub-sets t_{2A} of b_2 with the aid of the attributes $\mathfrak{V}(A)$ over **D**.

The general case is formulated analogously ($n > 1$):

(n.1′) $\bigwedge x(\mathsf{E}x\mathsf{b}_n \leftrightarrow \mathsf{E}x\mathsf{b}_{n-1} \vee x \subseteq \mathsf{b}_{n-1})$,

(n.2′) $\bigwedge x(\mathsf{E}x\mathsf{t}_{nA} \leftrightarrow \mathsf{E}x\mathsf{b}_n \wedge Ax)$.

For the sake of greater intelligibility we have here assumed that the attributes occurring on the right-hand side in the lines (..., 2′) are named, thus providing us with names (*viz* t_{nA}) for the sets occurring on the left-hand side. If we dropped this restriction we should have to replace the descriptions (n.2′) by mere existence formulas, *viz*

(n.2″) $\bigwedge A \bigvee z \bigwedge x(\mathsf{E}xz \leftrightarrow \mathsf{E}x\mathsf{b}_n \wedge Ax)$.

The conception that is given (temporary) expression in the above 'construction' and in the formulas describing it, is that any collections whatsoever of objects of our intuition or of our thought may be thought of as a whole, or: one object, that is: as a set. According to this conception, however, one should also be able to 'objectify', i.e. regard as forming a set, the totality of the sets introducible by the procedures so far described. This might require in addition a number of analogous procedures and all procedures, once accepted, may of course be applied in any order of sequence whatsoever. Whatever is thus introducible *should belong to* **D**.

One is tempted to express this by a formula

(a) $\bigwedge A \bigvee z \bigwedge x(\mathsf{E}xz \leftrightarrow Ax)$

which, interpreted for **D**, says that:

(a_1) Every property A over **D** defines a set z, *and*
(a_2) this z again belongs to **D**.[7]

However, this is clearly untenable. For a given **D** and a given interpretation of E, '$[x \mid \neg\, \mathsf{E}xx]$' describes a property A applicable to the objects in **D**. Substituting this specification, i.e. the property of a set x, of not being included among its own elements, in formula (a), we obtain

(b) $\bigvee z \bigwedge x(\mathsf{E}xz \leftrightarrow \neg\, \mathsf{E}xx)$.

Now assuming a z such that for *all* x $\mathsf{E}xz \leftrightarrow \neg\, \mathsf{E}xx$, it follows that, substituting z for x,

(c) $\mathsf{E}zz \leftrightarrow \neg\, \mathsf{E}zz$,

and this is a formula that is unsatisfiable in propositional logic, i.e. a contradiction.[8] It is known as *Russell*'s antimony, which demonstrates that the conception expressed in (a) is not tenable. There can thus be no maximal domain of sets in an 'intuitive' sense, but at most an open field of possibilities.

Such an open field, however, cannot be adequately described in the language of predicate logic, which must always refer to a constant, though arbitrarily selected, domain.

If we wish to retain the language of predicate logic, we shall have to impose restrictions. This can be done (α) in a purely pragmatic way, by imposing certain formal restrictions on the definition of the property A in (a) and therewith on the formula

(d) $\bigvee z \bigwedge x(\mathsf{E}xz \leftrightarrow \mathsf{A}(x))$

in the hope that this will exclude all contradictions and still leave sufficiently many of the intuitively meaningful procedures of set construction, cf. Quine [1], Essay V, and [2]; Ackermann [1];
(β) by limitation to certain procedures of set construction, which are explicitly formulated as special cases of (d), this latter not being accepted as generally valid.

From a formal point of view (β) is similar to (α), but since the former contains specifications of the accepted cases of (d), it has a certain constructive feature.

It is usually stipulated that **D** contains at least the sets described by (*n*.1) and (*n*.2) (or introducible in accordance with (*n*.1) and (*n*.2)), **B** being often assumed to be empty.[9] Let us suppose temporarily that **D** likewise consists of only the sets described by (*n*.1) and (*n*.2). Then every property that is meaningful for the **D**-things also determines a set in the intuitive sense. Certain of these sets already occur in **D** as objects, but new sets are certainly added as well.[10] Let **D*** be the domain that is formed out of **D** by the addition of these new sets. Let the interpretation of **B** remain as before and let the interpretation of E be correspondingly extended for the new elements of **D***.

Some formulas that are not satisfied by **D** now hold for **D***, e.g.:

(e) $$\bigvee z(\mathsf{E}\mathsf{b}_1 z \wedge \bigwedge xy(\mathsf{E}xz \wedge y \subseteq x \rightarrow \mathsf{E}yz) \wedge \bigwedge xy(\mathsf{E}xz \wedge \bigwedge v(\mathsf{E}vy \leftrightarrow \mathsf{E}vx \vee v \subseteq x) \rightarrow \mathsf{E}yz)),$$

for by its construction the set of all things out of **D** satisfies precisely the involved condition on *z*, and is also the only one in **D*** to do so. On the other hand, there are formulas that hold for **D** but not for **D***, e.g.:

(f) $\bigwedge x \bigvee y \mathsf{E}xy.$

The 'new' sets of **D*** are not elements of things *in* **D***. On the other hand, every thing in **D** is either an element of b_1 or a sub-set of some b_n. These, however, are elements of b_{n+1}.

The relation which is expressed here between the levels of **D** and **D*** is typical of 'construction' in set theory. Whenever a **D** has been established as totality of the sets introducible under the procedures already admitted, we may stop there; and in most cases the domains of sets thus described are beyond the average person's powers of imagination. However, one may consider it desirable to include this totality and its parts as new sets in a domain **D***, and to stop there for the time being.[11]

It then holds both 'for **D**' as well as 'for **D***' that

(g) $\bigwedge x \bigwedge A \bigvee z \bigwedge y(\mathsf{E}yz \leftrightarrow \mathsf{E}yx \wedge Ay)$

Further, the following variant of (a) holds, written here in an abbreviated form which we hope will be intelligible without strict definition:

(h) $\bigwedge A \bigvee z \bigwedge y(\mathsf{E}yz \leftrightarrow Ay)$, (i.e. A defined over **D**, z being in **D***).
D D* D

Not surprisingly a contradiction will result if the two levels **D** and **D*** are confused.

On the axiomatic basis suggested above we thus have the choice:

(A) To formulate all theorems for an indeterminate domain of sets to be subsequently fixed, though only closed under specified procedures. This gives us the advantage that ordinary predicate logic can be used; the disadvantage being that the admitted procedures must be formulated by axioms.

(B) To formulate all theorems for an *open* concept of sets. This has the advantage that no more or less arbitrary limit need be imposed; the disadvantage being that predicate logic, which is based on the conception of a fixed (though indeterminate) domain of objects, has to be modified. This can be done by a number of different available methods, and it is shown in the process that the most important concepts relating to language are open.

2. PROPOSITIONS THAT ASSERT THEIR OWN FALSEHOOD

The problem with which we shall be concerned in this section was already familiar to the Greeks and may be formulated as follows:

Someone (let us say: X) says:

(1) 'I am lying at this moment.'

Does X speak truly or falsely?

Or, to take another example, at 8 p.m. on July 6th 1968 someone makes the following remark in the course of a speech:

(2) 'What I say at 8 p.m. on July 6th 1968, is false.'

Or take the following non-fictitious example (cf. *Tarski* [2], p. 271; [3], p. 158):

(3) What is written in lines 6 and 7 on page 115 of this book, is false.

In all three cases we are concerned with a linguistic structure having the form of a proposition and making a statement about itself[12] (viz that it is false). This is important – unlike the objection that examples (1) and (2) cannot be regarded as instances of straightforward linguistic utterances being, in fact, reports about a linguistic utterance.

No difficulties seem to be raised by the notion of our talking in a language about this language, so long as we can distinguish clearly whether a statement is being made or whether something is being said about a statement. This means, however, that we must be able to name or describe propositions. For purposes of general discussion, we shall follow a convention established by Frege, whereby we use as name for a linguistic structure, this structure placed in inverted commas. Apart from this, we shall also use other names or descriptions, as e.g. in (1)–(3). Only this allows us to formulate a proposition that asserts something about itself, for under *Frege's* convention a proposition about the proposition A must always contain at least the inverted commas in addition to A,[13] i.e. must be longer than A.

If we define 'to lie' as 'to speak a falsehood', then the concept *false* occurs formally in each of the examples (1)–(3). If it is possible to represent this concept adequately in a language by means of a predicate 'is false', then all propositions must hold that are formed from the schema

(4) a *is false* if and only if it is not the case that A, by writing a proposition in place of 'A' and a name for this proposition in place of 'a', as in the following examples:

(4.1) '$3+2=5$' is false if and only if it is not the case that $3+2=5$,

(4.2) '$2 \cdot 2=5$' is false if and only if it is not the case that $2 \cdot 2=5$, and equally in cases where the name of the proposition is formed otherwise than under the *Frege*-convention.

Thus on the right-hand side of the above equivalences an assertion is made about numbers; on the left, about propositions about numbers.

Here, then, we have a counterpart to the requirement in respect of an adequate concept of truth, formulated by *Aristotle* and developed in detail by A. Tarski [2]:

(5) a *is true* if and only if A,

must yield a valid proposition demonstrable against an adequate definition, if 'A' is replaced by a proposition and 'a' by a description of this proposition.

Now if 'U' is an abbreviation and 'u' a name[14] for proposition (3), then on the basis of the construction of 'U' the following holds:

(6) u is false if and only if U;

and equally, applying (4) and (5):

(4.3) u is false if and only if not U,

(5.1) u is true if and only if U.

From (6) and (4.3) we now obtain (by virtue of the correspondence of the left-hand sides):

(7) U if and only if it is not the case that U;

and from (6) and (5.1) (by virtue of the correspondence of the right-hand sides):

(8) u is false if and only if u is true.

We shall limit our discussion to (7), since the formal contradiction contained in (8) is less easily demonstrated.

What, then, is the basis of the contradiction in (7)? On the assumption that the concept expressed by 'is false' has been meaningfully introduced, we have succeeded in formulating a proposition (viz the proposition on p. 115, lines 6, 7 of this book) that asserts its own falsehood. The concept 'is false' was regarded as a property of linguistic structures, defined by schema (4). However, the explanation of a new concept by means of a definition in the narrower sense, presupposes that the concepts used in the definition have been previously meaningfully introduced. This assumption is not met in the case of the application of the schema to proposition (3), as here the phrase 'is false', which has yet to be interpreted, occurs on the right-hand side. Of course, it is possible to explain the application of a new concept in stages, so to speak, and thus to obtain a definition in the wider sense: the application of the concept in one case is explained in terms of already explained cases (although eventually we have to return to cases where the definition is in the narrower sense). Such definitions always require to be specially validated. The above contradiction shows that definition (4) of 'is false' as well as definition (5) of 'is true' cannot be validated in this general form. The application of (4) to (3) results in a previously not explained case occurring on the right-hand side.

To obtain a genuine step-by-step definition, which can be validated, we proceed as follows: we first define the property expressed by 'is true', resp. by 'is false', for a part of the language[15] where these phrases are not used, then for the part for which the use of 'is true' and 'is false' is meaningfully explained by this definition, and so on. This procedure may be repeated any number of times, so that eventually definitions of truth, resp. of falsehood, are obtained for every proposition previously admitted into the language and containing one of these phrases.

However, by applying this procedure we do not, in fact, define two concepts *(true, false)* but two series of concepts, which may be designated more precisely by

'$true_1$', '$false_1$', '$true_2$', '$false_2$', '$true_3$', '$false_3$', ...

Now (3) cannot be formulated at all in this way.

There are comprehensive definitions of the form

(9) A is $true_\omega$ if A is $true_n$, where n is the smallest n for which 'A is $true_n$' is defined;[16]

(10) A is false_ω if A is false_n, where n is the smallest n for which 'A is $false_n$' is defined;[16]

but even such definitions do not enable us to extend the above procedure to cover the use of the phrases 'is true', 'is false' where no reference is made to stages, since the phrases 'A is $true_\omega$', 'A is $false_\omega$' are meaningful for propositions in which 'is $true_n$' or 'is $false_n$' occurs but not for propositions containing 'is $true_\omega$' or 'is $false_\omega$'. That is to say, even the concepts introduced by (9) and (10) only represent a segment. (The index ω, which we have employed in (9) and (10), is the customary set-theory symbol for the stage following the series 1, 2, 3, ...). The truth, resp. falsehood, of propositions containing '$true_\omega$', or '$false_\omega$' may then be defined in the subsequent stage, being symbolised by '$\text{true}_{\omega+1}$' and '$false_{\omega+1}$'.

Of course, we normally use the words 'true' and 'false' correctly without the addition of indexes: we have explained the use of such indexes here in order to draw the reader's attention to the segments involved in an exact definition of the open concepts '*true*', '*false*'; and it must be borne in mind that these segments are involved in any sound definition quite apart from the difficulties raised by the occurrence of the contradiction in (7).

On the other hand, if (4) and (5) are regarded not as definitions but as axioms for characterizing the concepts '*true*' and '*false*', then the contradition in (7) shows that there cannot be any concepts with these general properties. In the case of an axiomatic characterization of these concepts it is therefore just as necessary to make their openness explicit by the introduction of segments.

3. THE SET OF THINGS THAT CAN BE NAMED IN A LANGUAGE

The schema

(1) 'The set of things that can be named (in **L**)'

gives rise to antinomies, if we assume that the language **L** has certain possibilities of expression. For the sake of simplicity, let us allow the case where **L** is a somewhat artificially delimited part of a natural language (which, however, contains these possibilities of expression).

Let **L** be the totality of names, resp. of descriptions, of numbers in the English language, consisting of not more than one hundred letters.[17] Then the set *S* of natural numbers that can be named in **L** is in any case finite. For if, for the sake of simplicity and definiteness, we count the punctuation marks: full stop, comma, semicolon, as well as blanks, as letters, then we have in all 30 'letters'. Now if we imagine short names as made up with blanks to the length 100, then we can form $30^{100} = \underbrace{30 \cdot \ldots \cdot 30}_{100}$ sign sequences of length 100, of which only a part will be meaningful and only a sub-part descriptive of natural numbers (0, 1, 2, ...).

Since there are infinitely many natural numbers, there are numbers that cannot be named in **L** and among these, precisely one smallest one.

However, THE SMALLEST NATURAL NUMBER THAT CANNOT BE NAMED WITH ONE HUNDRED LETTERS, can be named with far fewer than one hundred letters, viz with 73, as simply counting will confirm. Now the limit ONE HUNDRED, which we chose for simplicity, could be refined if desirable.

For our present purposes, however, it suffices to state that:

The smallest natural number that cannot be named with one hundred letters, can be named with fewer than one hundred letters or, if we fill up with blank spaces, with one hundred letters.

How can this be explained? Let us call this curious number *X*. Now the very possibility of this short-hand description is odd, for now we have named the number with one single letter. However, in order to understand *this* designation we need to have a good deal of prior information, and this suggests that we investigate critically the concept of naming (more precisely, the relation: the word complex W *names* the thing z).

If we try to obtain an exact definition, we find that initially this is possible only for such W where the concept of *naming* does not occur, and we shall assume (analogously to VI 2) that we have thus defined 'W $names_1$ z'. On the basis of this definition we can define 'W $names_2$ z', where such W are admitted in which '$names_1$' occurs. And so on. It seems a fair assumption to make that a sequence of the above naming conventions (e.g.) '$names_1$', ..., '$names_8$' will yield a new naming convention '$names_9$' where the same things have in general considerably shorter names than previously. Our antinomy thus arose through oversight of the fact that the concept of naming is an open one.

We obtain an interesting variant of this antinomy if we apply its underlying schema to a language $\mathbf{L}_\Omega$ of the theory of so-called ordinal numbers. These are abstracted from the counting of segments of such iterations where we can meaningfully speak of the segment following upon an infinite sequence of segments.[18] This would be the case in our above example if 'W *names* z' were to be defined for such W where '$names_n$' is allowed to occur for all finite numerals *n* or even: where '$names_n$' with a variable *n* for finite numbers is used. Adopting the usual designation we should here write 'W $names_\omega$ z' instead of 'W *names* z'. Cf. also VI 1, VI 2.

As is customary, we have designated the first segment 'following' 1, 2, 3, ... with 'ω'. We then form $\omega+1$, $\omega+2$, $\omega+3$, ... whereupon follows $\omega+\omega$. Next we have $\omega+\omega+1$, $\omega+\omega+2$, $\omega+\omega+3$, ..., and then $\omega+\omega+\omega$. If '$\omega \cdot n$' is introduced as abbreviation for $\underbrace{\omega+\ldots+\omega}_{n}$, the subsequent series can be written more simply, viz: $\omega \cdot 3+1$, $\omega \cdot 3+2$, $\omega \cdot 3+3$, ..., which leads to $\omega \cdot 4$. The series $\omega \cdot 1$, $\omega \cdot 2$, $\omega \cdot 3$, ... is followed by $\omega \cdot \omega$, also written 'ω^2'. The series $\omega^2 \cdot 1, \omega^2 \cdot 2, \omega^2 \cdot 3, \ldots$ leads to $\omega^2 \cdot \omega$, i.e. ω^3; and the series $\omega^1, \omega^2, \omega^3, \ldots$ leads to ω^ω.

Now let S be a system of designations which, as suggested by our earlier examples, is to contain names for 'as many' ordinal numbers 'as possible', and where the names may be of any finite length whatsoever. Then 'the first ordinal number following the totality of ordinal numbers that can be finitely named in S' will be named by a *well-determined* ordinal number γ, if 'S' is replaced by a complete description of the system of designation indicated by 'S'. (For example: let S contain precisely what can be formed from 1, ω by the 'application' of $\alpha+\beta$, resp. additionally of $\alpha \cdot \beta$, resp. also of α^β.) This gives us an unambiguous description of γ *with reference to* the given system of designation S, but not *in* S. We reason analogously in the case of much more far-reaching systems of designation S; cf. in this connexion Bachmann [1] and, for a methodological refinement, Kleene [2].

On the other hand, the expression 'the smallest ordinal number that cannot be finitely designated, resp. named' would be a meaningful description of an ordinal number only if we could regard the totality of all finite systems of designation as *one* system of designation, which would then be *the* most comprehensive one. But precisely on this assumption we obtain a finite designation for the smallest ordinal number that cannot be finitely designated, and hence an antinomy.

Thus there can be no system of designation which is the most comprehensive, just as there can be no such domain of sets (VI 1) and no such definitions of truth and falsity (VI 2).

NOTES

1 Regarding *y* as an attribute we should say that *y* holds for *x*.

2 By way of specification we can either say: 'We all know what these are' or, more cautiously: 'They are chosen according to a purpose, but fundamentally the choice is arbitrary.'

3 Cf. Cantor [1]; translation taken from Fraenkel [1], p. 9.

4 Our initial use of the assumption is a cautious one, so as to make the subsequent misuse stand out more clearly.

5 It is thus assumed that for any set *s* whatsoever (i.e. in this context, $\mathfrak{b}_1$) we can constitute or 'think of' each of its sub-sets. In the case of infinite *s* this is a very large assumption which is tenable, if at all, only if we do not equate 'constituting' with 'describing' (by an individual definition). We have here avoided the term 'construct', since 'constructible' refers to a certain type of definability.

[6] By means of circumscriptions such as $\mathfrak{B}^*(\mathsf{b}_n) = \mathfrak{B}_n$, $\mathfrak{B}^*(A) = \mathfrak{A}$ we could, if we wished, avoid using a language whose correct *interpretability* has, after all, not yet been established.

[7] Because under the conventions of predicate logic the quantifiers $\bigvee z$ and $\bigwedge x$ *must* refer to the same domain of objects, i.e. in this case, **D**.

[8] It is sometimes claimed that this is no contradiction (which would have to be of the form $p \wedge \neg p$) but represents instead a kind of oscillation between the truth values T and F, since (c) contains the two implications $\mathsf{E}zz \rightarrow \neg \mathsf{E}zz$ and $\neg \mathsf{E}zz \rightarrow \mathsf{E}zz$. However, these will yield an immediate contradiction in the narrower sense *via* the propositional logic theorems $(p \rightarrow \neg p) \rightarrow \neg p$ and $(\neg p \rightarrow p) \rightarrow p$. Instead of the latter, $(\neg p \rightarrow p) \rightarrow \neg\neg p$ would be sufficient; this theorem is also valid in Intuitionist propositional logic, cf. IV 3, p. 79.

[9] Then b_1 is the empty set as element of **D**.

[10] In fact, 'very many', i.e. a number 'greater' than that of the elements of **D.**

[11] On the other hand, one may be interested in the sequence of these possible segments. The ordering of such levels is itself a subject-matter of set theory. We shall discuss this below in VI 3, p. 119 f., in terms of a somewhat simpler model.

Additionally we may observe: A concept as given by a formula in general changes its course-of-value, hence its meaning if that formula is interpreted (as) referring to different levels. In simpler cases the course-of-values in the extended model can be a continuation of the course-of-value of the 'shorter' model. This observation suggests certain 'identifications': concept = formula = course-of-value (the latter as the available part of something quite inexhaustible). But the hard fact that the continuability situation is restricted to fairly simple formulas should be a warning against the general constructivist identification of concepts and formulas.

[12] It might be objected that truth is not a property of linguistic structures but of their meanings. By transferring the problem to the linguistic level it can be more satisfactorily analysed; otherwise we are reduced to saying that (3) is meaningless.

[13] Thus if we had written: 'must always contain at least "A"', this would have meant 'must always contain at least the letter A'.

[14] Strictly speaking, the use of '(3)' as a name for the proposition under discussion is questionable, since such 'formula counters' are frequently regarded as abbreviations.

[15] Because of the indeterminateness of natural languages the procedure here outlined must be applied to a symbolic notation.

[16] And since an adequate definition of $true_{n+1}$ (resp. of $false_{n+1}$) comprises that of $true_n$ (resp. of $false_n$), we could write: where n is any n for which .... The index ω will be explained below.

[17] The choice of a limit in terms of numbers of words rather than letters might seem more obvious. However, as number words can easily be coalesced, the length of words would have to be restricted and this would complicate considerably our considerations below.

[18] Ordinal numbers are usually introduced within the framework of set-theory, where they first occurred (cf. for example Halmos [1]). A treatment of them as objects of a generalized arithmetic will be found e.g. in Bachmann [1]. Ordinal numbers are in a sense the prototype of an 'open totality'. For every given ('finished') set S of ordinal numbers there is a smallest ordinal number which is greater than all elements of S.

CHAPTER VII

LOGIC AND THE CRITIQUE OF REASON

The great questions of the critique of reason are: 'What can I know?', 'What ought I to do?', 'What may I hope?'.[1] Only to the first of these can logic provide a direct[2] answer and one, furthermore, that bears primarily on the knowledge systematized in the so-called *deductive* sciences. On the other hand, largely on the basis of investigations by Gödel, Tarski and others, we are today much better informed on this topic than could have been thought possible at the time when these questions were first formulated.

There exists a widespread misconception that what matters in mathematics, that prototype of all deductive sciences, is solely to have the right concepts and axioms, everything else – the working out of answers and decisions on problems through proof of relevant theorems – being only a matter of applying the appropriate rules of logic. It is held that mathematicians are guided solely by considerations of expediency or by aesthetic principles when deciding whether to tackle one problem rather than another; that they discover by trial and error which of the accepted rules to apply, and that at best, to help them with future problems, they develop a kind of 'sixth sense' – which, it is said, is precisely the quality that characterizes a good mathematician.

This idea would not be so far off the mark *if* the set of theorems valid for each field – delimited by a domain of formulas – *were* in fact defined by means of axioms and the accepted rules of inference, i.e. were defined syntactically in the sense of IV 2, p. 70. In fact, however, *validity* is usually, and in traditional mathematics always, defined – or at any rate must be assumed as being defined – *via* an *interpretation* in a determinate domain of objects, or in a well-determined totality of domains, as the case may be; i.e. it is defined semantically in the sense of III 3, p. 62, and a calculus is nothing else but a tool for the discovery of theorems. In many cases there are calculi which yield precisely all the originally semantically defined theorems (cf. IV 3), and this is probably the reason why many theories are given by a purely syntactic definition. There is a fairly obvious psychological motive here: in some vague sense a definition in terms of the tool used, seems more trustworthy. (Thus: 'Round means whatever can be produced on a turning-lathe.')

The above-mentioned results obtained by Gödel and Tarski and investigations based on them have, however, shown that there are problem complexes in mathematics – and therewith also in logic – that *cannot* be adequately treated by means of calculi. By a problem complex we here mean a domain of formulas **L** (e.g. in the form of III 2, C, p. 53 f.) and a semantic theorem or consequence definition for **L** (e.g. in the form of III 3, pp. 60, 61).

1. SETS THAT CAN BE PRODUCED COMBINATORIALLY. A GENERAL CALCULUS CONCEPT

If we disregard the fact that the expressions in the calculus definition in IV 2, p. 70 f., are interpretable, we are left with some rather curious rules for the production of sign strings. Now in order to obtain calculi that are as fruitful as possible, one will try to generalize the accepted form of the rules as far as is compatible with the minimum requirement, viz that the applicability of every basic rule must be verifiable in a finite number of steps. If we attempt to define 'verifiable' in this context by 'derivable in another calculus', we are faced with an endless regress, unless we stipulate limiting conditions for this latter calculus, a procedure which, however, is difficult to validate.

As our language for formulating rules, let us use the language $\mathbf{L}_F$ of predicate logic (cf. III 2, C, p. 52), or one of the extensions outlined in V, 1–3. Many more relations can be expressed in these languages than one will want to use for formulating rules, and it has been shown that all relations *whose holding-true cases are describable* by means of any one of the calculi[3] acceptable for this extended language, can already be expressed in $\mathbf{L}_F$. (It is, in fact, possible to avoid any explicit use of the concepts of *identity* (p. 51) and of *finiteness* (p. 103), intuitive as such use might be.)

We shall consider such interpretations **D** of $\mathbf{L}_F$ where the sign strings of the calculus *C* to be described belong to the domain of objects **D**, and where there occur as basic concepts: the combination ('concatenation') of sign strings, producibility in *C* ('provability'), and from case to case specific auxiliary concepts such as are required for the description of calculi.

Thus let C be a calculus with axioms and rules described by a formula A_C out of $\mathbf{L}_F$, where A_C comprises:

(1) a description of operations with the sign strings of C,

(2) a description of the axioms and rules of C.

Such a calculus C itself will describe a generally infinite catalogue C_1 in the sense of II 2, p. 31 for a property within the set of sign strings, for instance, formulas; and thus the formula A_C is also a 'description' of this catalogue $\mathbf{C}_1$. (The question arises whether a catalogue specified in any way whatsoever can be adequately described by a calculus.)

A description of the composition of C yields the formula A_C roughly thus:

Let the basic signs *(atoms)* of C be given in a specific sequence (e.g. $A, B, C, \ldots$ or $A_0, A_1, A_2, \ldots$). As series of names for these in the PC we then select the terms $a, f^1a, f^1f^1a, \ldots$ etc., formed with a specific object variable a and a specific function variable f^1.[4] We further select a two-place function variable f^2, which is to express the concatenation of sign strings. In this way we are able to describe all compound sign strings of C, and we choose a systematically distinguished[5] term $\mathtt{t}_Z$ designating the sign string Z. If B is the predicate with which the 'provability' of Z in C – *via* $B\mathtt{t}_Z$ – is to be expressed, then the rules of C are formulated with it and with additional auxiliary predicates. Thus A_C is now determined in principle.

The producibility of Z in the calculus C can be expressed *via* the demonstrability of $\mathsf{A}_C \rightarrow B\mathtt{t}_Z$ in the FC, i.e.

(3) $\vdash_F(\mathsf{A}_C \rightarrow B\mathtt{t}_Z)$ if and only if $\vdash_C Z$.

Let us say that the set of Z with the property that $\vdash_C Z$, is *regularly defined* by C, resp. by A_C. Since the $\mathbf{L}_F$-formulas are here used to say – *via* an interpretation – something about the calculus C, we should really write the semantic concept '$\models_F$' in place of '$\vdash_F$'. It is only by reason of the completeness of the FC that we can write '$\vdash_F$', and it is only by making this transition that we ensure that (3) does not 'cover too much'. (Think of $\models_{F2}$ instead of $\models_F$!)

Thus (3) places 'all thinkable' complexities of the rules of C within a formula A_C which is fixed for C, and from which they can be recovered by means of the rules of the FC. (In fact, as could be shown subsequently, only a very simple sub-section of the FC would be required.) It would thus seem to be convincingly demonstrated that in order to obtain a calculus that is to yield the most general combinatorially producible set, we need no more complicated rules than those of the FC.

We now define: a set S of sign strings is said to be producible combinatorially by means of rules, or *regular*[6], if there is a formula A so that for any sign string Z whatsoever out of the given store of symbols it holds that

(4) Z belongs to S if and only if $\vdash_F(\mathsf{A} \rightarrow B\mathsf{t}_Z)$.

These specifications are met, at any rate, by those sets S that are defined by a calculus C, i.e. that can be described by an A_C. But this is all, since the remaining A (as can be demonstrated *via* (5)) do not yield anything new.

Let us, for example, take the case where C is the FC itself. Admittedly, it would be a very laborious task to specify a formula A_{FC} with the property,that for every sign sequence Z of the FC it holds that:

(5) $\vdash_F(\mathsf{A}_{FC} \rightarrow B\mathsf{t}_Z)$ if and only if $\vdash_F Z$.

However, if we assume this to have been achieved[7], then we have the not very surprising result that the set of the theorems of the FC is regular. We should, after all, only have verified that the axioms and rules *for* the FC can also be formulated *in* the FC.

A. *A set that cannot be produced combinatorially*

Of much greater interest is the question whether there is also a formula U with which the set of non-theorems[8] of the FC can be described as regular, viz in terms of the condition

(6) $\vdash_F(\mathsf{U} \rightarrow B\mathsf{t}_Z)$ if and only if *not* $\vdash_F Z$,

for this would mean that we could give a positive characterization of the non-theorems in terms of the theorems.

Let us suppose, then, that there is such a U. We would then be able to characterize non-theorems *of the form* $(Z \rightarrow B\mathsf{t}_Z)$ in the same way, viz by a formula V together with

(7) $\vdash_F(\mathsf{V} \rightarrow Bt_Z)$ if and only if *not* $\vdash_F(Z \rightarrow Bt_Z)$.

Such a formula V could be constructed out of the assumed formula U e.g. as follows:

(8) $\mathsf{V} = \mathsf{U}(B/C) \wedge \mathsf{D}(A^2) \wedge \bigwedge y \bigwedge z(Cy \wedge A^2yz \rightarrow Bz)$.[8a]

If we now substitute the formula V for Z in (7), it will be seen that (7) and therewith also (6) are impossible. Thus the set of non-theorems of the FC cannot be regularly defined (Church's theorem [2]; the germ of a proof can be found already in Gödel [2]).

B. *On the generality of the approach*

By way of preparation for a discussion of the significance of this result, we shall present a number of arguments to show that our definition of regularity as a clarification of the intuitive concept of a set specifiable by rules of production, has the necessary generality. It is, at any rate, conceivable that 'more' could be obtained by replacing the FC in our definition by as powerful a calculus as possible for one of the extensions of predicate logic discussed in V 3. In reply to this objection we offer first this 'internal' argument: the FC is adequate for the description in the sense of (3) of all known calculi.

Over and above this, however, the following is an important 'external' argument: Several very different definitions have been proposed in an attempt to clarify the most general concept of producibility (or connected concepts), among them those of A. Church [1], K. Gödel [3], S. C. Kleene [1], A. A. Markov [1], A. Mostowski [1], E. L. Post [1], [2], R. M. Smullyan [1], A. M. Turing [1], (our definition (3) being an *apparent* generalization of Post [2] or Smullyan [1]). Many of these definitions are initially restricted to sets of natural numbers, and are then transferred with the aid of a constructive denumeration, so-called *Gödelization*, to domains of sign strings etc. Including this addition, where applicable, all definitions have so far shown themselves to be equivalent. This is a strong argument in support of the claim that each is right in its own way.

For a detailed treatment of the questions touched on in this section, the reader is referred to Davis [1] and Hermes [1].

2. DEGREES TO WHICH PROBLEM COMPLEXES ARE AMENABLE TO COMBINATORIAL TREATMENT

By a problem complex we here mean a domain of formulas, i.e. a well defined language **L** (cf. our remarks at the beginning of this chapter). If the definition of the theorems of **L** is already in the form of a decision procedure, as in the case of the semantic theorem definition for propositional logic in III 3, p. 57, or if it proves to be equivalent to such a definition, then in a sense we can speak of having complete mastery over the problem complex indicated by **L** – since every relevant question can be answered with the aid of the decision procedure, that is, we can recognize not only every theorem in **L** but can also uncover every non-theorem.

This 'recognition' can in all cases be given the form of an enumeration, or production, procedure (i.e. of a *calculus* in the sense of VII 1). It would not be difficult, though laborious, to express the construction of formulas of PC by means of an A_{C^+}, resp. A_{C^-} in the form of VII 1, (3), p. 124, with '+', resp. '−' indicating the cases of theorems, resp. non-theorems.

If, on the other hand, two calculi C^+ and C^- are given for a domain of formulas **L**, which enumerate two mutually complementary sub-sets of **L**, S^+ and S^- (in other words, S^+ and S^- have no elements in common and together constitute **L**), then these calculi together provide a procedure for deciding membership of S^+ (or of S^-): For every formula A of **L** the decision regarding membership of S^+ (and equally: of S^-) is obtained by denumerating all proofs of C^+ and C^-, alternating constantly between C^+ and C^-, until eventually A is 'judged'. If one were to restrict oneself to C^+, one would in general not know, so long as A was not caught by C^+, whether one had finished or not; and analogously for C^-. Thus C^+ and C^- together form a finite 'calculized' description of a catalogue C_2 in the sense of II 2, p. 32 in a domain **L** of formulas. The introduction of calculus C^- side by side with C^+ corresponds precisely to the transition discussed there from a catalogue $\mathbf{C}_1$ to a catalogue $\mathbf{C}_2$.

Let us say that S^+ is *coregularly defined* by C^- (similarly S^- by C^+), since S^+ may be described as 'the complement in **L**' of S^-, and since, *via* C^-, S^- may be said to be *regular* in the sense of VII 1. In general a set S is to be called *coregular* if the complement of S (in relation to a basic set **L**) is regular. The sub-sets S defined by a decision procedure in a set **L**, are thus both regular and coregular. Such sets shall therefore be called *biregular*.

This analysis of a decision procedure into two enumeration procedures, i.e. the two calculi C^+ and C^-, is in general not practically utilizable; nevertheless it is of theoretical interest, since it enables us to describe partial control of a problem complex.

Sets that are 'properly' (i.e. exclusively) regular or properly co-regular would appear to deserve equal interest in the sense of both being 'half-controllable'. If, however, we are concerned with the set **S** of the theorems of a theory θ, then there is a considerable difference: after a theorem P has been derived by means of C^+, and hence proved, we not only know something about the theorem as a formula, but we also know something about the things with which the theory θ deals. After deriving a non-theorem N through C^- – hence: after uncovering it – we in general know only that there is no sense in attempting a further derivation of N through C^+. We thus learn only indirectly about the things with which θ deals, since the negation of a non-theorem in general does not produce a theorem.

The conclusion reached in VII 1, A, p. 125 f., can thus be interpreted as follows: The set of theorems of the FC is properly regular, i.e. undecidable. Nevertheless, we are able to control the 'better half', since we have at our disposal an enumeration procedure for the theorems.

On the other hand, there are important problem complexes where the set of theorems is so defined that all non-theorems can be 'uncovered' by means of a systematic search for a counter-example – just as in the case of the PC – but where, diverging from the case of the PC, infinitely many trials would be required to determine theorems by means of the *definition*. Since the systematic search for counter-examples can be reduced to the form of a calculus, such sets of theorems are coregular, and there are very natural problem complexes where they are properly coregular. Thus no calculus acceptable for such sets of theorems can yield all theorems.

Over and above this there are problem complexes whose set of theorems is neither regular nor coregular – and among them are most of those which are obtained by restricting modern mathematical enquiries to appropriate formula domains **L**. This is even possible if **L** is the language of predicate logic, since the enquiries often lead to a theorem concept defined in terms of validity for a *specific* interpretation, or else for a *narrower totality* than was stipulated in the definition of the predicate-logic concept of validity. Such a problem complex cannot therefore be even half controlled. If, however, it is of such importance that even individual theorems are of interest, one will naturally try to find the most powerful calculi that are acceptable for the set of theorems. These calculi are then the only 'tangible' thing about the problem complex, and for this reason a problem complex is often identified with the set of theorems of its most powerful known calculus – particularly by researchers with constructivist leanings.

This procedure is adequate so long as one's aim is merely to demonstrate individual theorems, since for this purpose the best tool is the most powerful known calculus. On the other hand, the situation becomes more difficult if we attempt to extend to the general case the 'constructive definition'

true (resp. valid) = demonstrable,

which only happens to be adequate in the case of regular problem complexes. Such propositions A may then occur where neither A nor $\neg$A is 'true' (i.e. demonstrable). This is not surprising where calculi are concerned that are related to a totality of interpretations for the purpose of obtaining theorems; for A could be true for some of these interpretations and false for others. However, as will be shown by the examples in VII 3, p. 134, there are also non-regular sets of theorems which are defined in terms of a single interpretation. If one tried to maintain the 'constructive definition' for such cases, one would have to pay for the gain in precision by a loss in adequacy – and this is too high a price in my opinion.

3. PROBLEM COMPLEXES NOT AMENABLE TO COMBINATORIAL TREATMENT

The methods outlined above allow us to demonstrate that a problem *complex* is undecidable, that is, that there exists no common method for all problems of the 'complex' represented by a set of formulas. The same situation obtains if a problem complex is represented by *one* formula A with a variable a occurring in it, since this can represent, say, the set of sentences A(a/0), A(a/1), A(a/2), ...

The situation alters if we ask what is meant by saying that a specific individual problem, represented by a sentence A, is undecidable. If A belongs to a non-regular problem complex and if C is an admissible calculus for it, i.e. one that is necessarily incomplete, then A may be *undecidable in C*. This would be the case where A is a theorem of the problem complex not covered by C, so that A cannot be disproved either, because of the admissibility of C; in other words: A is not demonstrable. Under the conditions stated, there are thus *for* every C sentences undecidable *in C*.

Various attempts have been made to deal with this problem of the relativity of C. The suggested definition:

'undecidable' = 'undecidable in every admissible calculus'

is surely inadequate, since for every proposition A there is calculus C that is formally adequate to decide A: taking either of the sentences 'A' or '$\neg$A' as an additional axiom (under the usual rules), we always obtain one admissible calculus C. There is just no universal method that allows us to 'discover' such C. And such C could hardly count as a tool for determining the truth of A or of $\neg$A. This is by far less than we need; for 'absolute undecidability' ought to mean precisely that there can be no way to one particular case. And that is why there is no question here of referring to the most powerful known calculus. So far as I am aware, no one has yet succeeded in formulating an acceptable definition of the 'absolute undecidability' of individual problems.

Let us now look more closely at some typical undecidable problem *complexes*, i.e., using the terminology of VII 2, p. 128, non-biregular ones.

A. *Properly regular sets of theorems*

According to VII 1, p. 125 f., the set of theorems of predicate logic, or equivalently: of the FC, is properly regular, hence non-decidable. This, however, may be due to the very large expressive range of $\mathbf{L}_F$ (cf. III 2, C, p. 52 f.), and attention thus becomes focussed on reasonably delimited sub-sets of $\mathbf{L}_F$. On the one hand, the aim is to control as large sub-sets as possible through decision procedures.[9] On the other hand, there are difficulties, since already for quite small sub-sets S of $\mathbf{L}_F$ it is possible to show that the set of theorems thus limited to S remains properly regular. The proofs usually take the form of showing that the sub-set S already contains, though often very unobtrusively, the complete expressive range of $\mathbf{L}_{FC}$.

Thus, for example, already the set of theorems of the form $\mathsf{A}_{FC} \rightarrow Bt$ (cf. VII 1 (5), p. 125) is properly regular. The reason is that the whole complexity of the FC is here concentrated in the *one* formula A_{FC}, whilst Bt can express the demonstrability of any formula whatsoever.

Most of these proofs, however, have the form of one of the following examples:[10]

> Already the set of *theorems* of the form $\bigvee a_1 \ldots \bigvee a_n \mathsf{A}$, where A is constructed in terms of propositional logic out of atomic formulas *without* the use of $=$, but with function variables, is properly regular.
>
> Already the set of *theorems* of the form $\bigvee a_1 \bigvee a_2 \bigvee a_3 \bigwedge b \mathsf{A}$, where A is constructed in terms of propositional logic out of atomic formulas of $\mathbf{L}_F$ with at most two-place predicate variables and *without* function variables, is properly regular,

A further group of proofs concerns special mathematical theories θ, which can be formulated in $\mathbf{L}_F$ or $\mathbf{L}_I$. Let A_Θ be the formula formed by condensing the axioms of θ, and let B stand for any formula whatsoever with at most the same non-bound variables (the 'basic concepts of θ') as A_Θ, then the set of theorems of the form $\mathsf{A}_\Theta \rightarrow \mathsf{B}$ is properly regular for many important theories. For example, let us make the following substitutions among the $\mathbf{L}_I$-variables: let n stand for '0', e for '1', f^2 for '$[xy \mid x+y]$', g^2 for '$[xy \mid x \cdot y]$', A^2 for '$[xy \mid x<y]$'.[11] If we then formulate certain simple properties of the natural numbers (0, 1, 2, ...) and if these are condensed into a formula A_Θ, we obtain one of the simplest properly regular problem complexes.[12]

However, this problem complex is not 'the theory of numbers', since nothing need be said in A_Θ of the fact that the natural numbers consist of 'nothing more' than the series 0, 1, 2, ...[13]

B. *Properly coregular sets of theorems*

Whereas in the above examples the natural numbers were used merely as a background for an axiom system, properly coregular sets of theorems will be obtained if we define the theorem concept for specific formula domains $\mathbf{L}_i$ in terms of validity 'for the natural numbers'. The formula domains $\mathbf{L}_i$ may be thought of as sub-domains of $\mathbf{L}_I$; but for the sake of legibility we shall replace the variables n, e, f^2, g^2, A^2 (which here have a fixed interpretation as zero, one, addition, multiplication, smaller-than, i.e. as constants), by the customary 'constants': 0, 1, +, ·, <.

We shall extend a formula domain $\mathbf{L}_1$, which is obviously too narrow, step by step until the final extension results in a *properly coregular* set of theorems. That the sets of theorems occurring in the process are at any rate coregular, follows from the following property of all occurring formulas A: if all variables $x_1, \ldots, x_i$, on which depends the truth value of A, are replaced by numerals $\mathfrak{x}_1, \ldots, \mathfrak{x}_i$ – formally '$\mathsf{A}(x_1/\mathfrak{x}_1, \ldots, x_i/\mathfrak{x}_i)$' or simply '$\mathsf{A}(\mathfrak{x}_1, \ldots, \mathfrak{x}_i)$' – then the value of $\mathsf{A}(\mathfrak{x}_1, \ldots, \mathfrak{x}_i)$ can always be calculated in a finite number of steps.[14] But then we could in principle discover every non-theorem by calculating $\mathsf{A}(\mathfrak{x}_1, \ldots, \mathfrak{x}_i)$ in turn for all $\mathfrak{x}_1, \ldots, \mathfrak{x}_i$, where $\mathfrak{x}_1+\ldots+\mathfrak{x}_i=0$, $\mathfrak{x}_1+\ldots+\mathfrak{x}_i=1$, ..., $\mathfrak{x}_1+\ldots+\mathfrak{x}_i=k$. (In practice this procedure is, of course, out of the question in most cases.)

Let $\mathbf{L}_1$ consist of those equations $\mathsf{t}_1=\mathsf{t}_2$ that can be formed from terms constructed exclusively out of 0, 1, +, ·, x_1, x_2, ..., and parentheses (really: out of n, e, f^2, g^2, x_1, x_2, ...).

A well-known *theorem* formulable in $\mathbf{L}_1$ is

$$(1) \qquad x_1 \cdot (x_2 + x_3) = x_1 \cdot x_2 + x_1 \cdot x_3.$$

An example of a *non-theorem* is:

$$(2) \qquad x_1 + x_2 \cdot x_3 = (x_1 + x_2) \cdot (x_1 + x_3),$$

as is demonstrated by substituting $x_1/1$, $x_2/1$, $x_3/0$.

Elementary algebra provides an admissible calculus for the set of theorems on the natural numbers, formulable in $\mathbf{L}_1$. This set is thus bi-regular, and most of us learn to deal with it in the middle forms of our secondary school.

Let $\mathbf{L}_2$ consist of all formulas of the form $\mathsf{A}\rightarrow\mathsf{B}$, where A and B are $\mathbf{L}_1$-formulas. Although the value of $\mathbf{L}_2$-formulas can be calculated for given values of the variables just as easily as the value of $\mathbf{L}_1$-formulas – all we need to use additionally is the table for → (II 2, p. 37) – until 1969 we had no complete calculus for the set of $\mathbf{L}_2$-theorems nor a proof that this set is properly coregular. But see footnote 21.

Let $\mathbf{L}_3$ consist of all formulas that can be formed from $\mathbf{L}_1$-formulas by *any* propositional logic combination *whatsoever*. The problems formulable in $\mathbf{L}_3$ are, of course, generalizations of $\mathbf{L}_2$-problems; and, as can be demonstrated, they are inessential generalizations. For every $\mathbf{L}_3$-problem we can fairly easily find a finite set of $\mathbf{L}_2$-problems, whose solution would also solve the $\mathbf{L}_3$-problem in question.

Let $\mathbf{L}_4$ again be a generalization of $\mathbf{L}_3$: and let conjunctions and disjunctions of the form $\mathsf{A}(x/0) \wedge \mathsf{A}(x/1) \wedge \ldots \wedge \mathsf{A}(x/\mathtt{t})$ resp. $\mathsf{A}(x/0) \vee \mathsf{A}(x/1) \vee \ldots \vee \mathsf{A}(x/\mathtt{t})$ be additionally admitted, the term $\mathtt{t}$ in most cases containing variables; i.e. we are concerned with conjunctions and disjunctions with variable numbers of components. We could write instead in 'closed' form (i.e. without the reiteration dots)

$$\bigwedge_{x<\mathtt{t}+1} \mathsf{A}(x) \quad \text{resp.} \quad \bigvee_{x<\mathtt{t}+1} \mathsf{A}(x),$$

thus generalizing propositional logic —junctions; or we could write

$$\wedge x(x < \mathtt{t} + 1 \rightarrow \mathsf{A}(x)) \quad \text{resp.} \quad \vee x(x < \mathtt{t} + 1 \wedge \mathsf{A}(x)),$$ [15]

which formulas are rather particular examples of the expressive possibilities of predicate logic. It will be seen from the above motivation that the bound variable x must not occur in $\mathtt{t}$;[16] although, of course, variables y occurring in $\mathtt{t}$ may be bound by 'restricted quantifiers' $\wedge y(y<\mathtt{t}_1 \rightarrow \ldots)$ or $\vee y(y<\mathtt{t}_2 \wedge \ldots)$ placed further forward, as in the examples:[17]

(3) $\vee x(x < a + 1 \wedge \vee y(y < x + 1 \wedge a = x \cdot x + y \cdot y))$

(4) $\bigvee_{x<a+1} \bigvee_{y<x+1} \bigvee_{z<y+1} \bigvee_{u<z+1} (a = x^2 + y^2 + z^2 + u^2)$

As will be easily verified by trial and error, (3) is *not* a theorem of $\mathbf{L}_4$ (a counter-example is afforded by substituting '3' for a). On the other hand, (4) *is* a theorem of number theory[18] – a fact, however, that cannot be verified by trial and error. It cannot be verified directly, since this would require an infinite number of trials; nor can it be done in the wider sense in which we might say that by systematically listing all correct proofs for *one* suitable calculus, *every* theorem of number theory could be found, for

(5) the set of theorems on the natural numbers, formulable in $\mathbf{L}_4$ is properly coregular.

The proof, of which we cannot give details here, rests on the possibility of expressing the *non-demonstrability* of any calculus whatsoever in terms of the *validity* of suitable $\mathbf{L}_4$-formulas.

It is only *after* we have successfully formulated a proof that we know that a calculus – insofar as it is a systematization of specific methods of proof – is adequate for the solution of a problem posed by a formula, e.g. (4).

C. *Sets of theorems that are neither regular nor coregular*

Sets of theorems of this kind are obtained by means of certain straightforward extensions of the languages with which we have so far concerned ourselves. The point of view of predicate logic suggests extending $\mathbf{L}_3$ not into $\mathbf{L}_4$ but rather into a language $\mathbf{L}_5$, by admitting non-restricted quantifiers on natural numbers. Since every $\mathbf{L}_4$-formula A may be equivalently[19] replaced by an $\mathbf{L}_5$-sentence (viz by $\wedge x_1 \ldots \wedge x_n \mathsf{A}$, if $x_1, \ldots, x_n$ are precisely the variables still free in A), $\mathbf{L}_5$ lacks the property that characterizes[20] $\mathbf{L}_1$–$\mathbf{L}_4$, viz that the value of sentences is determinable in a finite number of steps. Moreover it is evident that:

(6) The set of $\mathbf{L}_5$-theorems is not regular.

For otherwise the set of $\mathbf{L}_5$-theorems of the form $\wedge x_1 \ldots \wedge x_n \mathsf{A}$, where A is a theorem of $\mathbf{L}_4$, and therewith the set of $\mathbf{L}_4$-theorems would be regular and also biregular in contradiction to (5).

(7) The set of $\mathbf{L}_5$-theorems is not coregular.

For otherwise the set of $\mathbf{L}_5$-theorems of the form $\neg \wedge x_1 \ldots \wedge x_n \mathsf{A}$ resp. $\vee x_1 \ldots \vee x_n \neg \mathsf{A}$, where A is a non-theorem of $\mathbf{L}_4$, and therewith also the set of $\mathbf{L}_4$-non-theorems would be coregular, and hence biregular in contradiction to (5).

Properties whose applicability to one number essentially involves the whole number series, can be formulated in $\mathbf{L}_5$. For example, the property of being the number of a formula which is undemonstrable in *C*, can be expressed for every calculus *C* through a formula *in* $\mathbf{L}_5$ – and not merely in metalogic, as would be the case if it were expressed *via* the *validity* of a formula of $\mathbf{L}_4$.

D. *Sets of theorems of logic*

The extensions of predicate logic itself, as explained and outlined in V 3, yield analogous results. The decisive step is the introduction of 'all' and 'there is (are)' for predicate variables, i.e. the addition of $\wedge \mathsf{A}^i$ and $\vee \mathsf{A}^i$ to the predicate logic symbolism; in other words the transition to $\mathbf{L}^2_{\mathrm{F}}$ in the sense of V 3, p. 101. The theorems formulated below for $\mathbf{L}^2_{\mathrm{F}}$ will also hold for all extensions of predicate logic that have at least the expressive range of $\mathbf{L}_\mathrm{F}^2$.

Since already the set of theorems of logic formulable in $\mathbf{L}_\mathrm{F}$ is not coregular, the same holds for all extensions of $\mathbf{L}_\mathrm{F}$ with analogously extended syntactic or semantic theorem definitions. The following holds for $\mathbf{L}_\mathrm{F}^2$ with a 'normal' semantic theorem definition (V 3, p. 101);

(8) The set of theorems of logic formulable in $\mathbf{L}_\mathrm{F}^2$ is not regular.

The demonstration of (8) is based on the possibility of using $\mathbf{L}_\mathrm{F}^2$-theorems to describe $\mathbf{L}_\mathrm{F}$-non-theorems, whose set is properly coregular, as shown in VII 1, p. 125 f. For the fact that such a $\mathbf{L}_\mathrm{F}$-formula A is a non-theorem, can be expressed *via* the general validity of a $\mathbf{L}_\mathrm{F}^2$-formula $\mathsf{N}(\mathsf{A})$ formed from A. If U is a $\mathbf{L}_\mathrm{F}^2$-sentence that asserts that there are infinitely many objects, and if $v_1, \ldots, v_k$ are all the variables occurring unbound in A, we can write:

(9) $\mathsf{N}(\mathsf{A}) =_{\mathrm{Df}} \mathsf{U} \rightarrow \neg \wedge v_1 \ldots \wedge v_k \mathsf{A}.$

Here we make use of the fact that for $\mathbf{L}_\mathrm{F}$-formulas validity in infinite domains coincides with general validity. (The general validity of the simpler formula $\neg \wedge v_1 \ldots \wedge v_k \mathsf{A}$ would express that *every* domain of objects provides a counterexample to A.)

There are so many $\mathbf{L}_\mathrm{F}^2$-sentences U_i that express the infinity of the domain of objects, that already the set of all propositions of the form $\mathsf{U}_i \leftrightarrow \mathsf{U}_j$ is neither regular nor coregular. The same holds for the set of propositions of the form $\neg \mathsf{U}_i \leftrightarrow \neg \mathsf{U}_j$. This transformation is indeed fairly simple; the result (Mostowski 1938), which may be formulated as follows, is all the more surprising:

> In no calculus can all possible definitions of finiteness be demonstrated as equivalent;

in other words:

> No codifiable system of deductive possibilities exhausts the meaning of the intuitive concept of finiteness.

NOTES

1 *Kant*, Critique of Pure Reason, Transcendental Doctrine of Method, p. 635 (A 805, B 833). Transl. by N. Kemp Smith, London 1929. A = 1st German edition, B = 2nd.

2 It can of course, provide an *indirect* answer to the other questions, too; e.g. I may not hope that at a specific time and in a specific place it will rain *and* not rain. Similarly in more important cases.

[3] This statement could be challenged, since it is necessarily based on our experience of tested rather than thinkable calculi. We assume, however, that any extension beyond the possibilities of FC would have shown itself within the tested calculi related to the languages outlined in V.
[4] If *C* only has a finite number (e.g. *n*) of atoms, then we can simply use the object variables $a_1 \ldots, a_n$ as their names.
[5] Thus if e.g. *A*, *B*, *C* are the atoms of *C*, and if *a*, *b*, *c* are the selected names in FC, then both f^2af^2bc and $f^2f^2\,abc$ are names for the sign string '*ABC*'. One is inclined to distinguish the simpler second form here, and in this case $f^2f^2f^2abbc$, for example, would be the distinguished name for '*ABBC*', i.e. t_{ABBC}.
[6] The standard term for this – referring to a different but equivalent definition – is 'recursively enumerable' (r.e.). Still other definitions are referred to by 'canonical' resp. 'formally representable' (f.r.): see Smullyan [1].
[7] Cf. the analogous definition in Scholz-Hasenjaeger [1], § 235.
[8] So far as the basic idea of the demonstration is concerned, it is immaterial whether we take these to be the non-theorems among the signstrings or among the formulas of the FC. In the case of a detailed proof this question would, of course, have to be settled.
[8a] That is, *B* is replaced throughout U by a 'new' *C*, and terms are then added that allow the transition from *C* to *B*. D (A^2) stands for the (somewhat complex) definition of an auxiliary concept A^2, which may be said to describe *y* as of the form $(z \rightarrow B\mathsf{t}_z)$.
[9] A report on the present state of these investigations will be found in Ackermann [2].
[10] Information about the present state of these enquiries will be found in Surányi [1].
[11] The comprehensors introduced formally in V 2, pp. 94 and 97, are here used to suggest the intended interpretations.
[12] This and many other results obtained in connexion with the form described in this section, are contained in Tarski-Mostowski-Robinson [1].
[13] As Th. Skolem [1] has shown, this fact cannot be expressed at all by means of axioms formulated in $\mathbf{L}_\mathrm{I}$.
[14] This is by no means self-evident: although only a finite number of words is required to define $\mathfrak{B}^*(\bigwedge x\mathsf{A})$, its calculation would require the infinitely many values $\mathfrak{B}^*(\mathsf{A}(0))$, $\mathfrak{B}^*(\mathsf{A}(1))$, $\mathfrak{B}^*(\mathsf{A}(2))$, ...
[15] We have written '$x < \mathbf{t} + 1$' to adapt the closed forms to our above formulations. The general form with '$x < \mathbf{t}$' would not result in properly increased expressiveness.
[16] $\bigwedge x(x < x + 1 \rightarrow \mathsf{A}(x))$ surely does not say the same as $\mathsf{A}(x/0) \wedge \mathsf{A}(x/1) \wedge \ldots \wedge \mathsf{A}(x/x)$; and analogously in the case of more complex terms.
[17] Where, for the sake of example, we shall use both the notations introduced above.
[18] Lagrange's theorem: Every natural number is the sum of four squares (*exactly* four if zero is admitted, otherwise *at most* four). The formulas express in addition that the squares are to be arranged in (weakly) decreasing order.
[19] In the sense of : $\models \mathsf{A}$ iff $\models \bigwedge x\mathsf{A}$, but in general not: $\models \mathsf{A} \leftrightarrow \bigwedge x\mathsf{A}$.
[20] At least: $\mathbf{L}_1$–$\mathbf{L}_4$ are typical instances; moreover, each extension of $\mathbf{L}_1$ or ... or $\mathbf{L}_4$ obtained by adjoining constants for computable functions and/or decidable attributes can be translated *into*, hence be understood *as a part* of $\mathbf{L}_4$.
[21] (Added in proof.) This instructive escalation has recently (1969, publ. 1970) been cut short by a result of Matiyasevich: With much harder efforts, $\mathbf{L}_4$ can even be translated into that sublanguage of $\mathbf{L}_2$, where B is a contradiction like $0 \neq 0$ or $0 = 1$.

CHAPTER VIII

TOWARDS THE LOGIC OF PROBABILITY

The theory of probability has a curious dual position: on the one hand, the validity of probability judgments forms a part of the subject matter of logic; on the other hand, its applications, for example, to games of chance, to mass phenomena and in particular in modern physics, often produce such complex combinations of the basic operations that a large part of the theory of probability consists in solutions to the resultant counting problems. The present introductory text cannot hope to deal with the mathematical techniques involved, techniques which in some cases yield merely approximate descriptions of the underlying conditions by means of an 'escape into infinity'. We shall, on the contrary, restrict ourselves to problems concerning the modes of validity of propositions, although occasionally we shall refer to results by way of example but without giving demonstrations. For the requisite mathematics the reader is referred to Jeffreys [1].

1. THE THEORY OF PROBABILITY AS A GENERALIZED SEMANTIC THEORY. A MEASURE OF POSSIBLE KNOWLEDGE

The two-valued semantic theory outlined in III 3 A, p. 56; C, p. 59; can readily be transferred schematically to arbitrary domains of 'values' for which counterparts to the truth functions are introduced, by means of which the proposition-forming expressions (cf. II 2, p. 36) are interpreted. Specific 'designated values' (one or more) will then correspond to the designated truth value **T**. The formal theory of such 'many-valued logics' has been considerably developed.[1] On the other hand, no interpretation of the values admitted in such logics as 'values representing validity' has been able to carry general conviction. Nor do 'mixed truth values' (in symbols e.g.: $x^2 \cdot \mathbf{T}+y^2 \cdot \mathbf{F}$, with $x^2+y^2=1$), as values of generalized attributes, appear to be suitable for describing imprecise concepts. Jan Łukasiewicz had earlier attempted to introduce one additional truth value **P** ('possible') between **T** and **F**; but although this has produced an interesting calculus, it is not one that sufficiently corresponds to the 'ordinary usage' of 'possible'. On the other hand, the outlook is more promising if we start from the *comparative* use of 'possible'. We can then introduce a whole scale of possibilities which may be described qualitatively, if imperfectly, e.g. as follows: *certain*, *almost certain*, *probable*, *possible*, *improbable*, *almost impossible*, *impossible*. The linearly ordered 'truth'-values of many-valued logic (see above) or the more complex 'truth'-table structures of modal logic[2] are intended as formal counterparts (i.e.: as clarifications) of these or such degrees. But I think the correspondence of the related calculi to the intuitive use of those concepts is not quite convincing (see Rosser–Turquette [1], pp. 3–8). On the other hand, a full scale of degrees of certainty (the appropriate calculus being some calculus of probability rather than syntactic many-valued or modal logic) is available from the so-called urn schema, as it yields a kind of 'standard measure' or 'weighting norm' for such degrees of possibility or propensities.[3]

A measure of the *certainty* of drawing a white ball at *one* random draw out of an urn containing black and white balls, which in all other respects are indistinguishable for practical purposes, is given by the ratio of the number w of white balls to the number b of black ones;[4] or better:[5] of w to $w+b$.

The fact that a *degree* of the *certainty* of their occurrence may be assigned as 'validity value' to specific propositions, viz propositions about the occurrence of possible events, suggests that such degrees of certainty be correlated to arbitrary formulas as validity values by means of generalized evaluations. The origin of these validity values will indicate which of the laws previously formulated (in III 3, p. 57) for evaluations, still hold. It goes without saying that, as in the case of $\{\mathbf{T}, \mathbf{F}\}$ assignments, we should not expect pure logic to determine the probability assignment 'valid' according to the state of affairs (described by a formula or by natural-language formulations).

If these probability assignments – which, in conformity with accepted usage in probability theory, we shall simply call *distributions* – are to be suitable for appropriately describing operations with degrees of certainty (possibly not restricted to propensities), then the laws formulated for them must at any rate accord with what we have learned from the example of the urns. For example, if A_1 and A_2 are propositions about urn-type experiments using different urns, i.e. independent experiments, but with $w{:}\ (w+s)=1/2$ for each, then the relevant part of the distribution $\mathfrak{W}$ is

$$\mathfrak{W}(\mathsf{A}_1) = \mathfrak{W}(\mathsf{A}_1 \wedge \mathsf{A}_1) = \mathfrak{W}(\mathsf{A}_2) = \mathfrak{W}(\mathsf{A}_2 \wedge \mathsf{A}_2) = 1/2,$$

but $$\mathfrak{W}(\mathsf{A}_1 \wedge \mathsf{A}_2) \ = \ 1/4,$$

for the case where $\mathsf{A} \wedge \mathsf{B}$ describes simultaneous experiments involving both urns. That is, $\mathfrak{W}(\mathsf{A} \wedge \mathsf{B})$ cannot be determined by $\mathfrak{W}(\mathsf{A})$ and $\mathfrak{W}(\mathsf{B})$ (cf. on the other hand, III 3, p. 57).

The ascription here of probabilities to formulas rather than to the events[6] (strictly: the types of events) they describe, requires to be justified. At any rate, formulas describing the same event, must be given the same value. This applies to 'logically equivalent' formulas and possibly to formulas equivalent under premises describing factual evidence. We suggest that the logic introduced in III 2, C and III 3, C be presupposed[7]; in the simplest cases, propositional logic (III 2, A and III 3, A) will be found sufficient.

A. *Distributions*

The following axioms V1–V4 are easily substantiated for formulas A, describing events with 'natural' propensities $\mathfrak{W}(\mathsf{A})$. We can, however, only give a partial answer here to the much more far-reaching question whether all probability assignments having the properties expressed by V1–V4, should be accepted as valid descriptions of degrees of certainty. (The outline of a more detailed analysis will be found in VIII 1, B, p. 145.) Basing ourselves on Kolmogorov [1], but omitting the (essential) part referring to infinite sums, we formulate:

V1. $\mathfrak{W}(\mathsf{A}) \geqq 0$
V2. If $\models \mathsf{A}$, then $\mathfrak{W}(\mathsf{A}) = 1$
V3. If $\models \mathsf{A} \rightarrow \mathsf{B}$, then $\mathfrak{W}(\mathsf{A}) \leqq \mathfrak{W}(\mathsf{B})$
V4. If $\models \neg(\mathsf{A} \wedge \mathsf{B})$, then $\mathfrak{W}(\mathsf{A} \vee \mathsf{B}) = \mathfrak{W}(\mathsf{A}) + \mathfrak{W}(\mathsf{B})$.

While V1 and V2 express a standardization of the scale of degrees of certainty; V3 expresses i.a. (see V8 below) that equivalent formulas also get equal certainty values. V4 however, describes a property of degrees of certainty that can easily be transferred from the urn example to arbitrary propensities, but not as obviously to degrees of certainty not based on propensities.[8] The question arises, what properties degrees of certainty (as abstract entities in their own right) must have for us to be able to 'measure' them in terms of numbers – like any other physical objects or states. (This would make the subsequent choice of the scale expressed in V1 and V2 merely a matter of convention.) It seems meaningful to assume:

(1) A linear order[9] with property V3 should be available for the degrees of certainty.
(2) If $\models \neg (\mathsf{A}_2 \wedge \mathsf{B}_2)$, $\mathfrak{W}(\mathsf{A}_1) \leqq \mathfrak{W}(\mathsf{A}_2)$, $\mathfrak{W}(\mathsf{B}_1) \leqq \mathfrak{W}(\mathsf{B}_2)$,
then also $\mathfrak{W}(\mathsf{A}_1 \vee \mathsf{B}_1) \leqq \mathfrak{W}(\mathsf{A}_2 \vee \mathsf{B}_2)$;

i.e., if both components in $\mathsf{A}_1 \vee \mathsf{B}_1$ are replaced by at least equally probable ones, we obtain an at least equally probable proposition, provided that the substituted formulas A_2, B_2 logically exclude each other.[10] It follows from (2) that on the assumption that $\models \neg(\mathsf{A} \wedge \mathsf{B})$, the value $\mathfrak{W}(\mathsf{A} \vee \mathsf{B})$ is uniquely determined by $\mathfrak{W}(\mathsf{A})$ and $\mathfrak{W}(\mathsf{B})$, let us say as $\Phi(\mathfrak{W}(\mathsf{A}), \mathfrak{W}(\mathsf{B}))$. There then follow from the logical properties of $\vee$ those properties of Φ, that allow $\Phi(g_1\ g_2)$ (for abstract degrees of certainty g_1, g_2) to be replaced precisely by $x+y$ (for numbers).[11]

Further properties of the distributions $\mathfrak{W}$ may now be deduced as theorems from axioms V1–V4. The following theorems have been selected so as to bring out the role of distributions as generalized evaluations $\mathfrak{B}^*$ (III 3 A). On the basis of V1–V4 it holds that:

V5. $\mathfrak{W}(\neg \mathsf{A}) = 1 - \mathfrak{W}(\mathsf{A})$ (from V1, V4)
V6. $\mathfrak{W}(\mathsf{A} \vee \mathsf{B}) = \mathfrak{W}(\mathsf{A}) + \mathfrak{W}(\mathsf{B}) - \mathfrak{W}(\mathsf{A} \wedge \mathsf{B})$
(from $\mathfrak{W}(\mathsf{A} \vee \mathsf{B}) = \mathfrak{W}(\mathsf{A}) + \mathfrak{W}(\neg \mathsf{A} \wedge \mathsf{B})$,
$\mathfrak{W}(\mathsf{B}) = \mathfrak{W}(\mathsf{A} \wedge \mathsf{B}) + \mathfrak{W}(\neg \mathsf{A} \wedge \mathsf{B})$)
V7. $\mathfrak{W}(\mathsf{A} \vee \mathsf{B}) \leqq \mathfrak{W}(\mathsf{A}) + \mathfrak{W}(\mathsf{B})$ (V6, V1)
V8. If $\models \mathsf{A} \leftrightarrow \mathsf{B}$, then $\mathfrak{W}(\mathsf{A}) = \mathfrak{W}(\mathsf{B})$ (V3)
V9. $\mathfrak{W}(\mathsf{A}) \leqq \mathfrak{W}(\mathsf{A} \vee \mathsf{B})$ (V3)
V10. $\mathfrak{W}(\mathsf{A} \wedge \mathsf{B}) \leqq \mathfrak{W}(\mathsf{A})$ (V3)
V11. $\mathfrak{W}(\mathsf{A} \wedge \mathsf{B}) \geqq \mathfrak{W}(\mathsf{A}) + \mathfrak{W}(\mathsf{B}) - 1$
V12. If $\models \mathsf{A} \wedge \mathsf{B} \rightarrow \mathsf{C}$,
then $\mathfrak{W}(\mathsf{A}) + \mathfrak{W}(\mathsf{B}) - 1 \leqq \mathfrak{W}(\mathsf{C})$ (V3 and V11)

For the case, assumed here for the sake of simplicity, that the 'world' can be described by a finite number of 'independent' propositions[12] $A_1, \ldots, A_n$ (e.g. by a catalogue $\mathbf{C}_2$ in the sense of II 2, p. 32) every complete description of a state of this world can be rendered in terms of a conjunction of the form

$$\text{(S):} \qquad [\neg]\,A_1 \wedge [\neg]\,A_2 \wedge \ldots \wedge [\neg]\,A_n,$$

where for every A the bracketed '$\neg$' may either stand or not. For n propositions A_i there are thus $2 \cdot \ldots \cdot 2 = 2^n$ propositions of form (S). A state s may also be given by that particular $\{\mathbf{T}, \mathbf{F}\}$ assignment $\mathfrak{B}_s$ of the 'variables' $A_1, \ldots, A_n$ (in the sense of III 3, p. 57) which satisfies the formula (S). We are here using the word 'independent' initially in the weaker sense in which every state S is possible, i.e., every $\mathfrak{B}_s$ is admitted.

Different conjunctions of the form (S) exclude each other, and every formula B constructed within propositional logic out of $A_1, \ldots, A_n$ is equivalent to a disjunction C constructed from specific formulas (S). Thus by repeated applications of V4 we obtain:

$$\text{V13.} \qquad \mathfrak{W}(\mathsf{B}) = \mathfrak{W}(\mathsf{C}) = \\ \ldots + \mathfrak{W}([\neg]\,A_1 \wedge [\neg]\,A_2 \wedge \ldots \wedge [\neg]\,A_n) + \ldots,$$

where 'on the right-hand side' a summand is yielded by precisely those formulas (S) from which B follows logically. Every distribution is thus[13] completely determined by the values for the 'state descriptions' (Carnap [1]). In other words: the distribution of 'weights' $g_{\mathfrak{B}}$ (whose sum is 1) on the 2^n possible assignments $\mathfrak{B}$, determines $\mathfrak{W}(\mathsf{B})$ as the sum of the $g_{\mathfrak{B}}$ for all assignments $\mathfrak{B}$ that satisfy B.

Since an assignment gives a complete description of the 'world', so that the 'correct' assignment $\mathfrak{B}$ corresponds to complete knowledge, a distribution may be understood as an expression of incomplete knowledge of the 'world':

$$\mathfrak{B}_1 \text{ has the probability } g_1, \ldots, \mathfrak{B}_{2^n} \text{ has the probability } g_{2^n}.$$

On the other hand, a formula B, which is equivalent to a disjunction C (with at least two components) formed out of state descriptions, also expresses incomplete knowledge.

At first sight these two kinds of incomplete knowledge appear to be incomparable; for in general B does not determine a $\mathfrak{W}$ (i.e., a sequence $g_1, \ldots, g_{2^n}$), and $\mathfrak{W}$ does not determine a B. But if, on the other hand, we regard sets[14] of distributions as expressions of possible knowledge, then the two can be reconciled. In this case $\mathfrak{W}$ is the set $\{\mathfrak{W}\}$ out of one distribution, whilst B comprises the set of all distributions $\mathfrak{W}$ with $\mathfrak{W}(\mathsf{B})=1$. (In other words: the states that are incompatible with B are weighted as 0, but otherwise weightings are left open.)

The distinguished distributions $\mathfrak{W}_{\mathfrak{B}}=\mathfrak{W}_{\mathfrak{B}_s}$ with

$$\mathfrak{W}_{\mathfrak{B}}(\mathsf{B}) = 1 \text{ if and only if } \mathfrak{B}^*(\mathsf{B}) = \mathbf{T},$$
$$\mathfrak{W}_{\mathfrak{B}}(\mathsf{B}) = 0 \text{ if and only if } \mathfrak{B}^*(\mathsf{B}) = \mathbf{F},$$

correspond to the value assignments $\mathfrak{B}$ (resp. to the evaluations $\mathfrak{B}^*$) discussed earlier in this section, and it therefore seems justifiable to regard distributions as generalized truth-value assignments.

Conversely, one might wish to extend the object language by introducing 'distribution descriptions' in a sense suggested by the following:

Let A, B be descriptions of different states (in the sense of p. 141) and let C, D be descriptions of distributions $\mathfrak{W}_{\mathsf{C}}$, $\mathfrak{W}_{\mathsf{D}}$, with e.g.

$$\mathfrak{W}_{\mathsf{C}}(\mathsf{A}) = 0.64, \quad \mathfrak{W}_{\mathsf{C}}(\mathsf{B}) = 0.36;$$
$$\mathfrak{W}_{\mathsf{D}}(\mathsf{A}) = 0.36, \quad \mathfrak{W}_{\mathsf{D}}(\mathsf{B}) = 0.64.$$

Since $\mathfrak{W}_{\mathsf{C}}(\mathsf{A} \vee \mathsf{B})=\mathfrak{W}_{\mathsf{D}}(\mathsf{A} \vee \mathsf{B})=1$, i.e., since $\mathsf{A} \vee \mathsf{B}$ also holds in the situation described by C (resp. by D), it seems fairly natural to extend the concept of consequence to the extent that for propositions like C, D it holds that

$$\mathsf{C} \models \mathsf{A} \vee \mathsf{B}, \quad \mathsf{D} \models \mathsf{A} \vee \mathsf{B},$$

and, additionally to use disjunctions like $\mathsf{C} \vee \mathsf{D}$ in such a way that we should also have

$$\mathsf{C} \vee \mathsf{D} \models \mathsf{A} \vee \mathsf{B}$$

(situations that are subsumed by $\mathsf{C} \vee \mathsf{D}$ – in the sense in which C is subsumed by $\mathsf{A} \vee \mathsf{B}$ – are also subsumed by $\mathsf{A} \vee \mathsf{B}$, if we already know what C, D, $\mathsf{C} \vee \mathsf{D}$ are).

In a discussion of questions such as whether it is always the case that

$$\models \mathsf{C}_1 \leftrightarrow \mathsf{C}_2 \text{ for } \mathfrak{W}_{C_1} = \mathfrak{W}_{C_2}$$

and whether (in the example) $\mathsf{A} \vee \mathsf{B} \models \mathsf{C} \vee \mathsf{D}$ also holds, i.e., *a fortiori:*

$$\mathsf{A} \models \mathsf{C} \vee \mathsf{D}, \quad \mathsf{B} \models \mathsf{C} \vee \mathsf{D},$$

it would turn out that not all laws of two-valued logic can be retained for a logic extended to cover propositions like C, D, $\mathsf{C} \vee \mathsf{D}$. For example, the so-called distributive law, the essential meaning of which may be expressed in

$$\mathsf{A} \wedge (\mathsf{B} \vee \mathsf{C}) \models (\mathsf{A} \wedge \mathsf{B}) \vee (\mathsf{A} \wedge \mathsf{C}),$$

would have to be abandoned.

The fact that $\mathsf{C} \models \mathsf{A} \vee \mathsf{B}$, resp. $\mathsf{D} \models \mathsf{A} \vee \mathsf{B}$, could tentatively be expressed, for example, by the following 'composition' of $\mathfrak{W}_C$ resp. $\mathfrak{W}_D$ out of $\mathfrak{W}_A$, $\mathfrak{W}_B$:

$$\mathfrak{W}_C = 0.64\, \mathfrak{W}_A + 0.36\, \mathfrak{W}_B,$$
$$\mathfrak{W}_D = 0.36\, \mathfrak{W}_A + 0.64\, \mathfrak{W}_B.$$

But this, by way of simple algebra, would uniquely determine the corresponding composition, (e.g.):

$$\mathfrak{W}_A = 2.2857\, \mathfrak{W}_C - 1.2858\, \mathfrak{W}_D,$$

and since this can hardly be interpreted as a mixture of possibilities, it is *not* in accordance with '$\mathsf{A} \models \mathsf{C} \vee \mathsf{D}$', which should remain acceptable in this context.

For comparison with the following we mention the correspondingly suggested definitions of C, D:

$$\mathsf{C} = 0.64\, \mathsf{A} + 0.36\, \mathsf{B}, \quad \mathsf{D} = 0.36\, \mathsf{A} + 0.64\, \mathsf{B},$$

which are subject to the same objections.

Anyway, any attempt to treat linear combinations of states should harmonize with the data of modern physics – which suggest an 'ontology of states'. A somewhat simplified ontology of this kind is expressed in the following definition. If $\mathsf{S}_1, \ldots, \mathsf{S}_k$ ($k=2^n$) are (resp. describe) the states (S) in the sense of p. 141, then the 'proposition'

$$\mathsf{Y} = x_1\mathsf{S}_1 + \ldots + x_k\mathsf{S}_k, \text{ with}^{15}\ x_1^2 + \ldots + x_k^2 = 1$$

shall describe the distribution $\mathfrak{W}_Y$ with[15] $\mathfrak{W}_Y(\mathsf{S}_i) = x_i^2$, ($i=1, \ldots, k$), so that (in terms of our example)

$$\mathsf{C}_1 = 0.8\, \mathsf{A} + 0.6\, \mathsf{B}, \quad \mathsf{D} = 0.6\, \mathsf{A} + 0.8\, \mathsf{B}$$

but equally e.g.

$$C_2 = 0.8\,A - 0.6\,B, \quad D \text{ as above.}$$

That on this definition we also have $A \models C_2 \vee D$ and $B \models C_2 \vee D$, follows from the conversions

$$A = 0.8\,C_2 + 0.6\,D, \quad B = -\,0.6\,C_2 + 0.8\,D,$$

which indicate that C_2, D and A, B have 'equal claims'. Whether the converses for C_1, D:

$$A = 2.8571\,C_1 - 2.1429\,D, \quad B = -2.1429\,C_1 + 2.8571\,D,$$

will similarly yield $A \models C_1 \vee D$, $B \models C_1 \vee D$ is, however, a question that cannot be answered without further analysis.[16]

The 'equality of claim' mentioned above suggests that all 'propositions' Y be regarded as state descriptions. This symmetry and the fact that different states (in our example: C_1, C_2) belong to the same distribution in respect of previously given states $S_1, \ldots, S_n$ (in our example: A, B), harmonizes so well with the data of quantum physics (data sometimes designated as paradoxical) that it may perhaps be desirable to base the extension of propositional logic outlined in this section on a footing as independent as possible of quantum physics. It seems feasible to suggest that a propositional logic expanded by means of distribution descriptions, will provide a framework within which the new data (of quantum physics) may be accommodated.[17]

It is questionable whether it is possible to establish a connexion between the $\mathfrak{W}_Y$ which we have here been considering, and the conditioned distributions $\mathfrak{W}_B$ (VIII 2, p. 149): from one point of view the $\mathfrak{W}_Y$ are more general, presupposing as they do an extended logic;[18] from a different point of view the $\mathfrak{W}_B$ are more general, because in them B is not restricted to state descriptions.

Without such a connexion, however, the above analysis does nothing to facilitate the important practical task of 'judging', on the basis of knowledge K, a proposition H not fully determined by K (i.e., neither $K \models H$ nor $K \models \neg H$, in terms of a *non*-expanded logic) by the allocation of a $\mathfrak{W}(H)$, – or, at any rate, an estimated $\mathfrak{W}(H)$. For, with the exception of the excluded limiting cases, all values for $\mathfrak{W}(H)$ are compatible with $\mathfrak{W}(K) = 1$.

The indication seems to be that we should try to express partial or complete ignorance not, after all, in terms of a set of distributions but through one suitably determined distribution. In the ideal case more and more instructive distributions would be obtained as a result of this initial knowledge being corrected by 'experience'.[19]

B. *A Measure of Possible Knowledge*

We conclude this section by discussing an attempt to characterize one initial or *a priori* distribution as an expression of minimal knowledge. This attempt is based on the following assumptions:

(1) The specific distributions $\mathfrak{W}_{\mathfrak{B}}$ contain maximal information about a 'world' describable by $\mathfrak{B}$.

(2) On the whole we find out *more*, if we learn which possibility out of a larger set of possibilities is the correct one.

This, however, needs to be qualified in several respects; among other things, the following must be made clear:

(3) We learn *more*, if the less probable of two possibilities proves to be the case.

(4) In the case of $m \cdot n$ 'equipossibilities' (which may be imagined as arranged in a rectangle consisting of m rows and n columns) the information consists of the information about the correct row and the information about the correct column.

A measure of information for $\mathfrak{W}$, which – abstracting from the content – measures only what we learn 'more' in the sense of (2) and (3), and which allows us in the case of (4) to *add* the measure for the rows to that of the columns, is given by the following definition of *Shannon's*:

$$(*) \qquad \mathbf{I}(\mathfrak{W}) = -(g_1 \cdot \log_2 g_1 + \ldots + g_k \cdot \log_2 g_k),^{20}$$

where $g_1, \ldots, g_k$ are the weightings correlated by $\mathfrak{W}$ to the possible k states resp. assignments. (*) comprises:

(a) If one $g_i = 1$ (i.e., if all others $= 0$), then $\mathbf{I}(\mathfrak{W}) = 0$.

(b) $-(\frac{1}{2} \cdot \log_2 \frac{1}{2} + \frac{1}{2} \cdot \log_2 \frac{1}{2}) = -\log_2 \frac{1}{2} = \log_2 2 = 1.$

That is, if $\mathfrak{W}(\mathsf{A}) = \mathfrak{W}(\neg \mathsf{A}) = \frac{1}{2}$, then we obtain the *unit of information* when we learn 'the truth' about A. This unit is known as a *bit* (an abbreviation of *binary digit*).

$$\text{(c)} \quad -\left(\frac{1}{2^n}\cdot\log_2\frac{1}{2^n}+\ldots+\frac{1}{2^n}\cdot\log_2\frac{1}{2^n}\right) = -\log_2\frac{1}{2^n}$$
$$= \log_2 2^n = n.$$

That is, if from among 2^n equipossible states (as described, e.g., by (S) on p. 141) we learn the correct one, then we have learned n bits. This corresponds to the description of (S) by n independent decisions about A_i. Here $k=2^n=1/g_i$. It can be shown that the above definition of $\mathbf{I}(\mathfrak{W})$ also characterizes arbitrary distributions:

(d) We learn $\log_2 1/g_i = -\log_2 g_i$ bits, if we learn that a state expected with a probability g_i actually obtains. Thus $\mathbf{I}(\mathfrak{W})$ is the average information gain (in bits) resulting from our learning the correct one of a number of states expected with the probabilities given by $\mathfrak{W}$. (Since the ith state with the probability g_i is the correct one, the 'weighted mean'[21] must here be formed with the weights g_i.)

Thus $\mathbf{I}(\mathfrak{W})$ is a measure of uncertainty (obtaining *before* any eventual complete information). This suggests that as a description of minimal knowledge we choose a distribution with maximum uncertainty. As can be demonstrated, this is uniquely determinable as the $\mathfrak{W}_0$ with $g_i=1/k$, in the case where k states are possible, i.e., an equal distribution.

This might be a satisfactory answer to our question *if* for every particular problem, we know the possible states from among which the one obtaining must be ascertained.

Let us take the case of an urn containing a known number of balls of two colours but with the ratio of the two colours unknown, the problem set being to discover the colour ratio by means of successive random draws, with each ball drawn being replaced before the next is drawn. Our reason for citing the urn example here, is that it provides a simple model for processes where (e.g. by reason of physical laws) we are *unable* to observe more. To be considered as possible state descriptions in this case are, e.g., the following:

(1) propositions that determine colour for every individual ball,
(2) propositions about the number of white balls,
(3) propositions expressing the occurrence of a specific colour.

It is evident that a requirement of equal distribution for (1), resp. (2), resp. (3) will in each case yield quite different distributions. If in this instance one thinks of (1) as the most natural choice, one introduces a kind of physics. This is inevitable if an infinity of competing possibilities is to be considered.

2. LOGICAL, SUBJECTIVE AND STATISTICAL PROBABILITY

The use of the urn schema as a weighting norm (VIII 1, p. 138) needs to be more precisely validated, and the starting point of such validation for degrees of certainty will differ from those for propensities, although eventually the resulting formulas will coincide. This is connected with the fact that axioms V1–V4 'hold' both for degrees of certainty as well as for propensities [22], despite the fact that their validations might differ. Here we must distinguish between on the one hand the structural content of the axioms, which expresses the capacity of the values for being ordered, the relation of this order to implication, and the existence of a function with the properties appropriate to disjunction [23]; and on the other hand, the conventional content which concerns simply the standardization of Φ (as sum) and therewith the choice of scale (from 0 to 1).

(a) If it is certain that a draw is being made, then it is certain that one of the $(w+b)$ balls will be drawn. Then the sum of the degrees of certainty for the propositions that each describe the finding of one specific ball, is 1. An equidistribution, describing minimal knowledge [24], will then assign to each ball (really: to the proposition that expresses its having been drawn) the probability $\frac{1}{w+b}$. Then the probability of an arbitrary white ball being found is given by V4 as $\frac{1}{w+b} + \ldots + \frac{1}{w+b} = \frac{w}{w+b}$. Probabilities calculated on the basis of a validation of this kind, may be termed logical probabilities, since they are grounded in a linguistic representation of ontological presuppositions (i.e., presuppositions concerning possible states). Cf. in this connexion Carnap [1], p. 162 ff.

(b) If for purposes of representing the knowledge contained in

$$\mathfrak{W}(\text{A draw will be made}) = 1,$$

we find ourselves confronted with the set of all such distributions $\mathfrak{W}$, then there still remains the possibility of selecting one admitted distribution in the expectation at best that this choice will be corrected by experience.[25] Here we should, of course, avoid 'sclerodox' prior judgments or prejudices, that cannot be corrected by any experience, i.e., we should not assign the weight 0 to any state that is still possible.[26] This interpretation of distributions is often called subjective probability (for a criticism see Carnap [1], p. 42 ff.). Perhaps it should be understood rather as a naive description or the correction of transformation of an *undetermined* 'choice' by that experience, or at its best: of a transformation of an undetermined initial *distribution of distributions*. (See p. 167 note 14.)

(c) If on the other hand we assume that the experimental conditions assign to every ball in the urn a propensity to be drawn – we may make this assumption on the basis of 'metaphysical' reasoning or within the framework of a physical theory – then perhaps the best way of getting beyond the general proposition $\mathfrak{W}(\ldots)=1$, is to infer the equality of the weights from the physical hypothesis of the 'symmetry of the experimental conditions'. In this case an equal distribution expresses the fact that the information is the maximal one available on the basis of the theory and the experimental conditions. We then have (in our example) $w : (w+b)$ as the propensity for 'white', i.e., formally as under (a).[27] The link between these interpretations is the assumption of equipossibility of competing events – though based in each case on different considerations. Let us refer in both cases to an assumption of *equipossibility of type 1*. Like every other physical hypothesis, that of symmetry, which includes in particular the assumption of the irrelevance of colour, is subject to the test of experience. Cf. in this connexion VIII 4, p. 162 f.

(d) We may reject as 'metaphysical' the assumption of the existence of propensities, e.g., for the behaviour of a real, that is, in general slightly unsymmetrical die: for example, the die may be destroyed after a small total number of throws (although this line of reasoning would also exclude the definition of probability as limiting value of relative frequencies). In this case we might attempt to express everything in terms of propositions about degrees of certainty. (Cases where the existence of an objective degree of certainty is doubtful, could be formally covered by stipulating that all values remain possible.)

Whereas degrees of certainty close to 1 ('almost certain') or close to 0 are intuitively accessible, other degrees may at first appear to be 'meaningless' since they say 'nothing' about individual cases. However, if we agree that there are cases where the same degree of confirmation p underlies each of a series of n 'independent' events, then a refinement of axioms V1–V4, which comprises the correction of distributions in the light of experience, enables us to assert 'almost certain' propositions about the 'relative frequency' of the occurrence of the event in question. For: with a 'large' n it is 'almost certain' that the ratio of cases of occurrence to the total number n is 'in the neighbourhood of' p. This will be clarified below, but first a word of caution. On an entirely different basis an assumption of *equipossibility – of type 2*, as distinct from the one introduced earlier – will here yield a relation between a probability and a ratio of 'favourable' to 'possible' cases. But the difference in validation alone should warn us not to 'define' 'the' concept of probability in terms of such a ratio.

(e) Basing ourselves on a frequency interpretation we may define the probability of A on the presupposition of B,[28] viz $\mathfrak{W}_B(\mathsf{A})$, out of $\mathfrak{W}$ by means of

$$(*) \qquad \mathfrak{W}_B(\mathsf{A}) = \frac{\mathfrak{W}(\mathsf{A} \wedge \mathsf{B})}{\mathfrak{W}(\mathsf{B})},$$

We thus introduce $\mathfrak{W}_B$ as the correction, conditional on experience B, of $\mathfrak{W}$. It readily follows that several corrections may be comprised together, i.e.:

$$(\mathfrak{W}_B)_C(\mathsf{A}) = \frac{\mathfrak{W}_B(\mathsf{A} \wedge \mathsf{C})}{\mathfrak{W}_B(\mathsf{C})} = \frac{\mathfrak{W}(\mathsf{A} \wedge \mathsf{C} \wedge \mathsf{B}) \cdot \mathfrak{W}(\mathsf{B})}{\mathfrak{W}(\mathsf{B}) \cdot \mathfrak{W}(\mathsf{B} \wedge \mathsf{C})} = \frac{\mathfrak{W}(\mathsf{A} \wedge \mathsf{B} \wedge \mathsf{C})}{\mathfrak{W}(\mathsf{B} \wedge \mathsf{C})}$$

$$= \mathfrak{W}_{B \wedge C}(\mathsf{A}), \quad \text{in brief:} \quad (\mathfrak{W}_B)_C = \mathfrak{W}_{B \wedge C}.$$

Despite the additional 'knowledge' B, $\mathfrak{W}_B$ can represent greater ignorance than $\mathfrak{W}$, cf. for example VIII 4, p. 164.

(f) Let us now find a different basis for (*), since we do not wish to take the correspondence of degrees to frequency phenomena for granted. We shall express the fact of dependence on the experience expressed in the 'evidence' B, by means of a second argument – thus recognizing that in general a probability judgment depends on the evidence available. Let us symbolize these evaluations, which have again been generalized, by $\mathfrak{w}(\mathsf{A}, \mathsf{B})$ – for $\mathfrak{W}_B(\mathsf{A})$. For a constant knowledge B we then obtain as a counterpart to V1–V4, axioms W1, W2, W3, W4. These will be supplemented by a readily intuitive counterpart W3′ to V8, which expresses that what is 'logically equivalent' is also of equal value as evidence (cf. Carnap [1], p. 285).

W1. $\mathfrak{w}(A, B) \geqq 0$

W2. If $C \models A$, then $\mathfrak{w}(A, C) = 1$

W3. If $C \models A \rightarrow B$, then $\mathfrak{w}(A, C) \leqq \mathfrak{w}(B, C)$

W3′. If $\models B \leftrightarrow C$, then $\mathfrak{w}(A, B) = \mathfrak{w}(A, C)$

W4. If $C \models \neg(A \wedge B)$, then $\mathfrak{w}(A \vee B, C) = \mathfrak{w}(A, C) + \mathfrak{w}(B, C)$.

W2, W3, W4 have here been strengthened as against V2, V3, V4 in that in each case a presupposition of 'logical truth' is replaced by one of 'factual truth' (i.e., on the basis of the respective non-contradictory evidence C).

We now add an axiom that relates distributions on the basis of different evidence:

W5. If $A \models C$, $B \models C$, $A \models D$, and $B \models D$,

then $\dfrac{\mathfrak{w}(A, C)}{\mathfrak{w}(B, C)} = \dfrac{\mathfrak{w}(A, D)}{\mathfrak{w}(B, D)}$ [29]

W5 expresses that the ratio of degrees of certainty (from A to B) is independent of any change of evidence (C resp. D) as far as only consequences of A, and of B separately (i.e.: of $A \vee B$) are considered. The particular choice $D = A \vee B$ could be used to simplify the axiom, but the chosen version has the advantage of being as free of particular concepts as possible.

Derivations from V1–V4 (cf. p. 140 f.) may be transferred in an analogous sense. We merely note for subsequent use

W6. If $C \models A \leftrightarrow B$, then $\mathfrak{w}(A, C) = \mathfrak{w}(B, C)$ (cf. V8).

The most important derivation, which is based essentially on W5 and which is usually formulated as an axiom,[30] is

W7. $\mathfrak{w}(A \wedge B, C) = \mathfrak{w}(A, C) \cdot \mathfrak{w}(B, A \wedge C)$

Proof. Propositional logic yields

(1) $C \models A \wedge B \leftrightarrow A \wedge B \wedge C$,

(2) $C \models A \leftrightarrow A \wedge C$,

(3) $A \wedge B \wedge C \models C,$

(4) $A \wedge B \wedge C \models A \wedge C,$

(5) $A \wedge C \models C,$

(6) $A \wedge C \models A \wedge C,$

(7) $A \wedge C \models A \wedge B \wedge C \leftrightarrow B.$

Suitably substituting, we then derive from W6 with (1), (2)

$$\text{(8)} \qquad \frac{\mathfrak{w}(A \wedge B, C)}{\mathfrak{w}(A, C)} = \frac{\mathfrak{w}(A \wedge B \wedge C, C)}{\mathfrak{w}(A \wedge C, C)}$$

Further, from W5 with (3), (4), (5), (6)

$$\text{(9)} \qquad = \frac{\mathfrak{w}(A \wedge B \wedge C, A \wedge C)}{\mathfrak{w}(A \wedge C, A \wedge C)}$$

And from W6 with (7) and (6)

$$\text{(10)} \qquad \frac{\mathfrak{w}(A \wedge B, C)}{\mathfrak{w}(A, C)} = \frac{\mathfrak{w}(A, A \wedge C)}{1} = \mathfrak{w}(A, A \wedge C)$$

Hence by cancelling out the denominator, W7. In the form (10), W7 is a precise counterpart to the definition of $\mathfrak{W}_B(A)$ discussed above under (e). Important derivations[31] from W7 are

$$\text{W8.} \qquad \mathfrak{w}(A, B \wedge C) = \mathfrak{w}(A, C) \cdot \frac{\mathfrak{w}(B, A \wedge C)}{\mathfrak{w}(B, C)}$$

and

$$\text{W9.} \qquad \frac{\mathfrak{w}(A_1, B \wedge C)}{\mathfrak{w}(A_2, B \wedge C)} = \frac{\mathfrak{w}(A_1, C)}{\mathfrak{w}(A_2, C)} \cdot \frac{\mathfrak{w}(B, A_1 \wedge C)}{\mathfrak{w}(B, A_2 \wedge C)}$$

Proofs. Because of $\models B \wedge A \leftrightarrow A \wedge B$, for line (2), we have

(1) $\mathfrak{w}(B, C) \cdot \mathfrak{w}(A, B \wedge C) = \mathfrak{w}(B \wedge A, C)$ (W7)

(2) $= \mathfrak{w}(A \wedge B, C)$ (W6)

(3) $= \mathfrak{w}(A, C) \cdot \mathfrak{w}(B, A \wedge C)$ (W7)

Dividing by $\mathfrak{w}(B, C)$, we then obtain W8. And, substituting A_1 resp. A_2 for A in W8 and dividing, W9 follows as an immediate inference, since $\mathfrak{w}(B, C)$ which is independent of A is eliminated.

(g) We may fairly assume that one and the same degree of certainty underlies a sequence of events (see above under (d)), in the case where random draws are made out of an urn, with each ball drawn being replaced before the next is drawn, or in the case where random throws are made with an unbiassed die (if we are justified in assuming that the die is not altered by being thrown). (Whether propensities come into play, shall here be left open, since our aim is to understand the situation in terms of degrees of certainty.) If C expresses our general prior knowledge, A the outcome of the preceding draws resp. throws[32], B the observational result about to be obtained, then according to the assumptions implicit in the experiment, we have

$$\mathfrak{w}(\mathsf{B}, \mathsf{A} \wedge \mathsf{C}) = \mathfrak{w}(\mathsf{B}, \mathsf{C}) \quad (\text{briefly: } = p),$$

i.e., the preceding observations give no additional information about the outcome of the next attempt. If $\mathsf{B} = \mathsf{B}_n$ refers to the nth attempt, then we may have

$$\mathsf{A}_{n-1} = [\neg]\,\mathsf{B}_1 \wedge \ldots \wedge [\neg]\,\mathsf{B}_{n-1}$$

i.e., in general either

$$\mathsf{A}_n = \mathsf{A}_{n-1} \wedge \mathsf{B}_n \quad \text{or} \quad \mathsf{A}_n = \mathsf{A}_{n-1} \wedge \neg\, \mathsf{B}_n$$

If we now apply W7 in order to determine $\mathfrak{w}(\mathsf{A}_n, \mathsf{C})$, we obtain *either*

$$\begin{aligned}\mathfrak{w}(\mathsf{A}_n, \mathsf{C}) &= \mathfrak{w}(\mathsf{A}_{n-1} \wedge \mathsf{B}_n, \mathsf{C}) = \mathfrak{w}(\mathsf{A}_{n-1}, \mathsf{C}) \cdot \mathfrak{w}(\mathsf{B}_n, \mathsf{A}_{n-1} \wedge \mathsf{C}) \\ &= \mathfrak{w}(\mathsf{A}_{n-1}, \mathsf{C}) \cdot \mathfrak{w}(\mathsf{B}_n, \mathsf{C}) = p \cdot \mathfrak{w}(\mathsf{A}_{n-1}, \mathsf{C})\end{aligned}$$

or

$$\begin{aligned}\mathfrak{w}(\mathsf{A}_n, \mathsf{C}) &= \mathfrak{w}(\mathsf{A}_{n-1} \wedge \neg\, \mathsf{B}_n, \mathsf{C}) \\ &= \mathfrak{w}(\mathsf{A}_{n-1}, \mathsf{C}) \cdot \mathfrak{w}(\neg\, \mathsf{B}_n, \mathsf{A}_{n-1} \wedge \mathsf{C}) \\ &= \mathfrak{w}(\mathsf{A}_{n-1}, \mathsf{C}) \cdot \mathfrak{w}(\neg\, \mathsf{B}_n, \mathsf{C}) \\ &= (1 - p) \cdot \mathfrak{w}(\mathsf{A}_{n-1}, \mathsf{C}).\end{aligned}$$

For a sequence of n observations with results described by A_n and with B and $\neg\mathsf{B}$ occurring g and $(n-g)$ times, respectively, we obtain the following permutational analysis:[33]

$$\mathfrak{w}(\mathsf{A}_n, \mathsf{C}) = p^g \cdot (1 - p)^{n-g}.$$

Let us begin by grouping together all possible observational sequences with the same g[34] and then all sequences with $g:n$ close to p. (All sequences that are different at one point mutually exclude each other, so that their weights can be added together.) The precise formula for the degree of certainty of $g:n$ falling between $p-\varepsilon$ and $p+\varepsilon$ (where ε is thus a measure of imprecision) is so unwieldy, that in most cases we use an approximating formula – which, however, yields a precise estimate of the degree of certainty. We thus obtain a relationship between the imprecision (ε), the number of attempts (n) and the degree of confirmation $\mathfrak{w}(p-\varepsilon \leqq g:n \leqq p+\varepsilon, \mathsf{C})$, where '$p-\varepsilon \leqq g:n \leqq p+\varepsilon$' indicates a disjunction of all propositions about trial sequences of length n with $g:n$ between $p-\varepsilon$ and $p+\varepsilon$. Let us give an example. For an urn with balls of six different colours in equal numbers or for a 'good' die, $p=1:6=0.1\overline{6}=0.1666\ldots$. Then for a sequence of 1000 trials we may expect the result $g:n$ as follows:[35]

between	with the (degree of) certainty s	$s:(1-s)$[36]
$p \pm 0.008$	0.503	1.0
$p \pm 0.01$	0.604	1.3
$p \pm 0.02$	0.910	10
$p \pm 0.03$	0.989	90
$p \pm 0.04$	0.999 31	1 500
$p \pm 0.05$	0.999 978	45 000
$p \pm 0.06$	0.999 999 64	2 800 000
$p \pm 0.07$	0.999 999 997 1	340 000 000
$p \pm 0.08$	0.999 999 999 988	860 000 000 000

In order to double the precision obtained (which is the same as halving the limits for $g:n$) with the *same degree of certainty*, four times as many trials are required; and for k times the precision, k^2 as many. On the other hand, the table below (with p again $= 1:6$) shows how the degree of certainty increases with increasing length n of the trial series if the precision is constant ($\pm$ 0.008 in our example).

Length n	certainty s	$s:(1-s)$
1 000	0.5	1
1 560	0.6	1.5
2 310	0.7	2.3
3 590	0.8	4
6 250	0.91	10
14 400	0.99	$100=10^2$
23 600	$0.999=1-10^{-3}$	10^3
32 800	$1-10^{-4}$	10^4
42 300	$1-10^{-5}$	10^5
52 000	$1-10^{-6}$	10^6
81 400	$1-10^{-9}$	10^9
approx. 100 000	$1-10^{-11}$	10^{11}

It should be noted that accordingly there is no length n, for a prescribed degree of precision, that guarantees with absolute certainty that this precision will be met, although on the other hand such sequences yield values $g:n$ 'in the neighbourhood of' p not only 'in the long run' but in general already within their fairly long segments. This corresponds exactly to our experiences with sequences of trials incorporating randomizing devices. A *definition* of probability as limit of a sequence of relative frequencies (von Mises [1], p. 17) does not seem to me to take due account of these experiences, although it has been possible to avoid the formal contradictions of von Mises' original formulation.

We have calculated predictions with a 'high degree of certainty' about the behaviour of observable sequences, basing ourselves on the general properties of degrees of certainty, and assuming an equipossibility of kind 2 together with independence of the trials one from the other – which is basically equivalent to assuming that we have as datum an object of the theory restricted to frequency phenomena.[37] On the other hand, even those who wish to restrict the application of the concept of probability to frequency phenomena, will ultimately have to ascribe a degree of certainty to that individual event which consists in the total sequence having a specific property (viz its relative frequency being within a given interval).[38] This again seems to suggest that degrees of certainty, i.e., types of validity, be regarded as *the* objects of the theory of probability – a conception that would still allow us to base assumptions of equipossibility of kind 2 made in applications of the theory, on special assumptions of similarity (e.g., about the presence of propensities). In this way statistical probability may be subsumed as a special case under the concept of degrees of certainty.

The above-mentioned connexion between p and $g:n$ indicates that the determination of $g:n$ from a sufficiently long series of observations be regarded as a measurement of p, understood as the (same) degree underlying each case. An initial difficulty arises from the fact that the requisite propositions about the precision and certainty of this precision would presuppose knowledge of p. However, this may be circumvented by exploiting a different connexion between $g:n$ and p, viz one that is based essentially on W8 resp. on W9. Cf. in this connexion VIII 4, p. 165.

3. RULES OF INDUCTIVE INFERENCE

From the point of view of traditional logic, inductive inference is one from the particular to the general – in contrast to deductive inferences, for which the inference 'from the general to the particular' (which is represented in our symbolism by '$\bigwedge x \mathsf{A}(x) \models \mathsf{A}(y)$') is regarded as a particularly characteristic example. Since what one has in mind here are, of course, 'reasonable' inductive inferences, the word 'particular' is used to mean a body of experience admittedly incomplete yet nonetheless sufficiently large to allow general laws to be 'inferred' – as (apparently) happens successfully in the empirical sciences. Thus we might say that the decisive factor is the drawing of inferences from incomplete information; and that it is plausible that such inferences carry degrees of (un)certainty.

The importance of 'inductive' inferences arises from the fact that all information yielded by observations on a sufficiently 'rich' world is incomplete. We cannot, however, establish their *validity* by arguing that they have proved themselves so far and will therefore continue to do so. For this would be to argue in a circle, since our reasoning would be based on an inference of the kind to be validated. However, one might regard such reasoning as the abstract form of a behaviour pattern innate in man – and presumably also in animals[39] – and one could then infer from the fact that such beings (still) exist, that behaviour on this pattern is appropriate to our environment, at any rate partially. Our environment therefore, so the argument would run, at least approximates to the ideal structure, on the basis of which inductive inferences are justified.

However, we shall no doubt have to accept the fact (as in the case of two-valued logic[40]) that the only way of avoiding a vicious circle is to undertake an analysis rather than a validation. And here again it may turn out that an adequate analysis of a theory θ presupposes a theory θ^* more powerful than θ.

In view of the fact that inductive inferences are often made unconsciously[41] – with a greater or lesser degree of skill – let us begin by drawing attention to the underlying rules. Some of these rules are given in Pólya [1], [2], at first as rules of plausible inference in a qualitative formulation,[42] and include:

$$\text{P1.}\quad \frac{\begin{array}{l} A \text{ implies } B \text{ (i.e.: } B \text{ follows from } A\text{)} \\ B \text{ (turns out to be) true} \end{array}}{A \text{ is more credible, or likelier.}}$$

$$\text{P2.}\quad \frac{\begin{array}{l} A \text{ implies } B \\ B \text{ is credible} \end{array}}{A \text{ is (somewhat) more credible, or likelier.}}$$

$$\text{P3.}\quad \frac{\begin{array}{l} A \text{ implies } B \\ B \text{ is very improbable in itself} \\ B \text{ is (however) true} \end{array}}{A \text{ is very much more credible.}}$$

In these general formulations the comparatives are still, so to speak, 'in the air'. Perhaps an example in illustration of P3 will indicate what needs to be added in every instantiating case: If $\mathfrak{a}$ is a poisoner, then $\mathfrak{a}$ must have procured poison. It is very improbable that anyone should buy poison. However $\mathfrak{a}$ has bought poison. Consequently: it is much more likely (than before this information was obtained) that $\mathfrak{a}$ is a poisoner.

One is tempted to describe the subjective formulations here, such as 'credible', by introducing degrees of certainty. However, Pólya himself prefers to think of his rules as pointers to the discovery of mathematical theorems, and it seems doubtful to me whether $\mathfrak{w}(\ldots, \ldots)$ resp. $\mathfrak{W}(\ldots)$ could be meaningfully applied to mathematical statements. (Can a probability alter as a result of a successful mathematical demonstration?) On the other hand, the objection that unique chance occurrences cannot be thus described, could be met by pointing out that in such cases the theory yields only relations *between* degrees of certainty, and not values *of* them.

If by rules of inductive inference we understand rules where probability statements occur in a premise or in the conclusion, then the simplest such rules are those that form a link between deductive and inductive inferences. The following holds:

$$\text{R1.}\qquad \frac{\begin{matrix}\mathsf{K}, \mathsf{A}_1, \mathsf{A}_2 \vDash \mathsf{B} \\ \mathfrak{w}(\mathsf{A}_1, \mathsf{K}) \geqq 1 - \varepsilon_1 \quad \mathfrak{w}(\mathsf{A}_2, \mathsf{K}) \geqq 1 - \varepsilon_2\end{matrix}}{\mathfrak{w}(\mathsf{B}, \mathsf{K}) \geqq 1 - (\varepsilon_1 + \varepsilon_2)}$$

where K expresses the knowledge available and where probabilities close to 1 are indicated by $1-\varepsilon$ (i.e., with small 'uncertainty' ε).

Proof. From $\mathsf{K}, \mathsf{A}_1, \mathsf{A}_2 \vDash \mathsf{B}$, it follows that

(1) $\mathsf{K} \vDash \mathsf{A}_1 \wedge \mathsf{A}_2 \rightarrow \mathsf{B}$, hence, with W3

(2) $\mathfrak{w}(\mathsf{A}_1 \wedge \mathsf{A}_2, \mathsf{K}) \leqq \mathfrak{w}(\mathsf{B}, \mathsf{K})$.

On the other hand, from W1–W4 there follows the counterpart to V11

(3) $\mathfrak{w}(\mathsf{A}_1 \wedge \mathsf{A}_2, \mathsf{K}) \geqq \mathfrak{w}(\mathsf{A}_1, \mathsf{K}) + \mathfrak{w}(\mathsf{A}_2, \mathsf{K}) - 1$.

On the basis of the presuppositions of R1, we have

(4) $\mathfrak{w}(\mathsf{A}_1, \mathsf{K}) + \mathfrak{w}(\mathsf{A}_2, \mathsf{K}) - 1 \geqq 1 - \varepsilon_1 + 1 - \varepsilon_2 - 1$
$= 1 - (\varepsilon_1 + \varepsilon_2)$,

hence

(5) $\mathfrak{w}(\mathsf{A}_1 \wedge \mathsf{A}_2, \mathsf{K}) \geqq 1 - (\varepsilon_1 + \varepsilon_2)$.

Finally, the assertion of R1 follows from (2) and (5). The following generalization of R1 is proved similarly:

$$\text{R1*.}\qquad \frac{\begin{matrix}\mathsf{K}, \mathsf{A}_1, \ldots, \mathsf{A}_n \vDash \mathsf{B} \\ \mathfrak{w}(\mathsf{A}_1, \mathsf{K}) \geqq 1 - \varepsilon_1, \ldots, \mathfrak{w}(\mathsf{A}_n, \mathsf{K}) \geqq 1 - \varepsilon_n\end{matrix}}{\mathfrak{w}(\mathsf{B}, \mathsf{K}) \geqq 1 - (\varepsilon_1 + \ldots + \varepsilon_n)}$$

Because of $\mathsf{K}, \mathsf{A}, (\mathsf{A} \rightarrow \mathsf{B}) \vDash \mathsf{B}$, an initial application of R1 yields the rule

R2. $$\frac{\mathfrak{w}(\mathsf{A}, \mathsf{K}) \geqq 1 - \varepsilon_1 \quad \mathfrak{w}(\mathsf{A} \to \mathsf{B}, \mathsf{K}) \geqq 1 - \varepsilon_2}{\mathfrak{w}(\mathsf{B}, \mathsf{K}) \geqq 1 - (\varepsilon_1 + \varepsilon_2)}$$

In the same way there corresponds to every deductive step in a deductive argument, a probability inference with accumulating degrees of non-confirmation.

In the case of an application of R1* the certainty attainable by B does not depend on the complexity of a derivation of B out of $\mathsf{A}_1, \ldots, \mathsf{A}_n$; but in the case of a sequence of inferences of kind R2, all degrees of uncertainty accumulate. Thus in general one will fare worse, i.e., obtain a weaker conclusion, if instead of applying R1* *once* at the end of a purely deductive proof, one applies the probability inference analogous to R2 at *every* stage. Perhaps we may see in this a justification for a logic that is more precise than the conditions to which it is to be applied.[43]

The following modes of inference may be regarded as counterparts to certain plausibility inferences, although translation into the language of degrees of certainty is not straightforward:

R3. $$\frac{\mathsf{K}, \mathsf{A} \models \mathsf{B}}{\mathfrak{w}(\mathsf{A}, \mathsf{K} \wedge \mathsf{B}) \geqq \mathfrak{w}(\mathsf{A}, \mathsf{K})}$$ (cf. P1)

The premise in P1: '*B* is true' is thus taken into account in that the probabilities 'for K' and 'for $\mathsf{K} \wedge \mathsf{B}$' are compared. It should be noted that here – as in P1 – the inference 'from B to A' involves a kind of *reversal* of the deductive premise.

A refinement of R3 is

R4. $$\frac{\mathsf{K}, \mathsf{A} \models \mathsf{B} \quad \mathfrak{w}(\mathsf{B}, \mathsf{K}) < 1 \quad \mathfrak{w}(\mathsf{A}, \mathsf{K}) > 0}{\mathfrak{w}(\mathsf{A}, \mathsf{K} \wedge \mathsf{B}) > \mathfrak{w}(\mathsf{A}, \mathsf{K})}$$

Proofs for R3 and R4. The premise $\mathsf{K}, \mathsf{A} \models \mathsf{B}$ yields that $\mathsf{K}, \mathsf{A} \models \mathsf{K} \wedge \mathsf{A} \leftrightarrow \mathsf{B}$. Therefore according to W2 and W6

(1) $\mathfrak{w}(\mathsf{B}, \mathsf{K} \wedge \mathsf{A}) = 1.$

According to W8 we have

(2) $\mathfrak{w}(\mathsf{A}, \mathsf{K} \wedge \mathsf{B}) \cdot \mathfrak{w}(\mathsf{B}, \mathsf{K}) = \mathfrak{w}(\mathsf{B}, \mathsf{K} \wedge \mathsf{A}) \cdot \mathfrak{w}(\mathsf{A}, \mathsf{K}).$

Hence, with q for $\mathfrak{w}(\mathsf{B}, \mathsf{K})$ and with (1):

(3) $\mathfrak{w}(\mathsf{A}, \mathsf{K} \wedge \mathsf{B}) \cdot q = \mathfrak{w}(\mathsf{A}, \mathsf{K})$

i.e., for $q \neq 0$

(4) $\mathfrak{w}(A, K \wedge B) = (1/q) \cdot \mathfrak{w}(A, K)$

Because of $K \models B \rightarrow K$ we have $q \leqq 1$ (cf. W2 and W3), i.e., $1/q \geqq 1$. Then the assertion of R3 follows from (3). For the strengthened assertion R4 we require $q < 1$ (which corresponds to the second premise in P3) and also $\mathfrak{w}(A, K) > 0$, since otherwise everything in (3) could be $= 0$. For very small q (but with $q \neq 0$) $1/q$ is very large. This gives us a variant of R4 that is still closer to P3.

The presupposition that B does not follow from K alone[44] is expressed in a different way by the following variant of R4:

R5. $$\frac{K, A \models B \quad \mathfrak{w}(B, K \wedge \neg A) < 1 \quad 0 < \mathfrak{w}(A, K) < 1}{\mathfrak{w}(A, K \wedge B) > \mathfrak{w}(A, K)}$$

Proof. Because of $K \models B \leftrightarrow (B \wedge A) \vee (B \wedge \neg A)$, we have

(1) $\mathfrak{w}(B, K) = \mathfrak{w}((A \wedge B) \vee (\neg A \wedge B), K)$

(2) $= \mathfrak{w}((A \wedge B, K) + \mathfrak{w}(\neg A \wedge B, K)$ (W4)

(3) $= \mathfrak{w}(A, K) \cdot \mathfrak{w}(B, K \wedge A) +$
$\mathfrak{w}(\neg A, K) \cdot \mathfrak{w}(B, K \wedge \neg A)$ (W7)

Because of $K, A \models B$ we have $\mathfrak{w}(B, K \wedge A) = 1$, i.e.,

(4) $\mathfrak{w}(B, K) = \mathfrak{w}(A, K) + \mathfrak{w}(\neg A, K) \cdot \mathfrak{w}(B, K \wedge \neg A)$

(5) $= \mathfrak{w}(A, K) +$
$\mathfrak{w}(\neg A, K) \cdot (1 - (1 - \mathfrak{w}(B, K \wedge \neg A)))$.

(6) $= 1 - \mathfrak{w}(\neg A, K) \cdot (1 - \mathfrak{w}(B, K \wedge \neg A))$

(7) $= 1 - (1 - \mathfrak{w}(A, K)) \cdot (1 - \mathfrak{w}(B, K \wedge \neg A))$.

According to the premises of R5 the product on the right-hand side is not 0, i.e., $\mathfrak{w}(B, K) < 1$; thus R4 is applicable.

By formulating rules R3, R4, R5 after the pattern of P1 and P3 we have perhaps veiled the essential meaning of these inferences. It may be clearer in the following formulation:

R6. $$\frac{0 < \mathfrak{w}(B, K \wedge \neg A) < \mathfrak{w}(B, K \wedge A) \quad 0 < \mathfrak{w}(A, K) < 1}{\mathfrak{w}(A, K \wedge B) > \mathfrak{w}(A, K)}$$

Proof. In view of W9 (with A for A_1, $\neg A$ for A_2)

$$(1)\qquad \frac{\mathfrak{w}(A, K \wedge B)}{\mathfrak{w}(\neg A, K \wedge B)} = \frac{\mathfrak{w}(A, K)}{\mathfrak{w}(\neg A, K)} \cdot \frac{\mathfrak{w}(B, K \wedge A)}{\mathfrak{w}(B, K \wedge \neg A)}$$

According to the premises both right-hand quotients are meaningful, and therefore we have $\dfrac{\mathfrak{w}(B, K \wedge A)}{\mathfrak{w}(B, K \wedge \neg A)} > 1$, i.e.,

$$(2)\qquad \frac{\mathfrak{w}(A, K \wedge B)}{\mathfrak{w}(\neg A, K \wedge B)} > \frac{\mathfrak{w}(A, K)}{\mathfrak{w}(\neg A, K)}$$

Since here numerator + denominator $= 1$ in each case, we have

$$(3)\qquad \mathfrak{w}(A, K \wedge B) > \mathfrak{w}(A, K).$$

Let us assume, by way of example, that we know (K) that an urn which we have is one of two with different ratios of 'black' and 'white'. Let A describe the case of ours being the 'whiter' urn, and let B stand for a white ball being drawn (under the usual conditions of drawing and replacing). Then the premises of R6 will have been met, and the conclusion expresses the plausible fact that every 'white' draw increases the probability of our urn being the 'whiter' one.

Our example also enables us to compare R6 with rules R3–R5. For the sake of simplicity, we shall not change the number of urns involved. Then R3 corresponds to the borderline case where all that is known is that the urn described by A is 'pure white'; on the other hand, the application of R5 requires the additional knowledge that the urn described by $\neg A$ is 'mixed'. And R4 is the more obvious rule to apply if a (specific) value for $\mathfrak{w}(B, K)$ is known even though this is not expressed by '$\mathfrak{w}(B, K) < 1$'.

4. PROBABILITY AND TRUTH. ON OUR DEPENDENCE ON A PRIOR JUDGMENT

Rules R3–R6 do not directly contribute to the solution of the practically important tasks of inductive logic, which are

(A) to make probability statements about a hypothesis H (i.e. to calculate a degree of certainty $\mathfrak{w}(H, K)$), on the basis of knowledge expressed in the truth of a proposition K;

(B) to decide on the truth of a proposition on the basis of the knowledge of relative frequencies (as approximating values for degrees of certainty).

Thus such inferences always depend on additional presuppositions about distributions, and these presuppositions must therefore be validated 'in some way or other'.

Further to (A): When we determine $\mathfrak{w}(\mathsf{H}, \mathsf{K})$ resp. $\mathfrak{W}_1(\mathsf{H})$ with

$$\mathfrak{W}_1 = [\mathsf{H} \mid \mathfrak{w}(\mathsf{H}, \mathsf{K})] \qquad \text{(cf. V 2, p. 97)}$$

we in effect distinguish one distribution as an expression of logical probability; or at any rate, we restrict the domain of admissible distributions by means of objective criteria, since except for borderline cases all values for $\mathfrak{W}_1(\mathsf{H})$ are compatible with the presupposition $\mathfrak{W}_1(\mathsf{K}) = \mathfrak{w}(\mathsf{K}, \mathsf{K}) = 1$. However, the indeterminateness of the $\mathfrak{W}_1$ must not be understood as the determinateness of a $\mathfrak{W}_\mathsf{X}$ through X according to the formula[45]

$$\mathfrak{W}_\mathsf{X}(\mathsf{H}) = \mathfrak{w}(\mathsf{H}, \mathsf{X}) = \frac{\mathfrak{W}_0(\mathsf{H} \wedge \mathsf{X})}{\mathfrak{W}_0(\mathsf{X})} \qquad \text{(cf. VIII 2, p. 149)}$$

with an already distinguished $\mathfrak{W}_0$, since all available knowledge – i.e., including any expressed in X – is comprised in K. In fact, all $\mathfrak{W}_0$ with $\mathfrak{W}_0(\mathsf{X}) \neq 0$, are still to be taken into account here.

In the light of V 2 (p. 97), we have $\mathfrak{W}_0 = \mathfrak{W}_\mathsf{L} = [\mathsf{H} \mid \mathfrak{w}(\mathsf{H}, \mathsf{L})]$ with 'L' standing for an arbitrary propositional logic theorem. It thus seems more appropriate (rather than attempt *ad hoc* determinations of $\mathfrak{W}_1$) to distinguish, if possible, *one* $\mathfrak{W}_0$ as an expression of logical probability, thus also determining $\mathfrak{w}$.

It has turned out that the assumption of equipossibility of kind 1 suggested by the propositional logic structure of the object language, is not always appropriate;[46] but that, e.g., the structure of monadic predicate logic indicates other symmetries and hence equipossibilities of kind 1. This has led R. Carnap (cf. Carnap [2]) to develop methods for the determination of $\mathfrak{W}_0$. These have so far allowed $\mathfrak{W}_0$ to be specified for monadic predicate logic (which in any case is essentially more comprehensive than traditional syllogistic), the specification depending on a decision regarding the extent to which 'items of *a priori* knowledge' resp. 'empirical facts' are to influence a judgment.

Thus we may select a distribution $\mathfrak{W}_0$, but all values calculated from it will reflect the bias already inherent in our *choice of language*, which selects a finitely describable part out of the wealth of phenomena.

Further to (A) and (B). If we make an assumption of equipossibility of kind 1 (including, where appropriate, in the above-mentioned extended sense) in a *sequence* of cases, then this amounts to an assumption of equipossibility of kind 2, and Bernoulli's analysis (VIII 2 (g), p. 152 f.) may then be applied on the basis of the inferred equipossibilities of kind 2.

We may then test the substantiation of $\mathfrak{W}_0$ as follows. The degree of confirmation p calculated from $\mathfrak{W}_0$ for the 'same' event $\mathfrak{r}$ (in the cases under consideration) is set against the observed relative frequency $g:m$ for the occurrence of $\mathfrak{r}$:

If $\mathfrak{W}_0$ has been correctly determined, then

$g:m$ is almost certainly close to p;

i.e., since our antecedent here does not depend on the events:

It is *almost certain* that for a correctly determined $\mathfrak{W}_0$

$g:m$ is *close to p*,

i.e., if the observed ratio $g:m$ is not *close to p*,

It is *almost certain* that $\mathfrak{W}_0$ (together with its substantiation) is not appropriate.

Apart from making clear what we mean by *close to* and *almost certain* (cf. VIII 2 (g), p. 153), we also need to decide on the degree of certainty that is to bridge the gap between the almost certain and the true – in order that the above considerations provide an example of a solution of task (B), viz to reach a (substantiated) decision on $\mathfrak{W}_0$ on the basis of $g:m$.

But even so, our example only provides a partial solution of task (B), for it permits at most a negative judgment (other $\mathfrak{W}_0$ could yield the same p). On the other hand, we did not require any additional presuppositions about distributions.

No doubt the main problem involved with tasks of type (B) is to render such presuppositions harmless:

The example of the two urns (p. 160) may serve as a (highly simplified) model for the task of deciding, on the basis of observations, on the 'correctness' of theories making predictions about observable events differing from each other only in degree of certainty (of happening). Thus let A_1 express that we have an urn with 2/3 white and 1/3 black balls, and A_2 that we have one with 1/3 white and 2/3 black balls.[47] Further, let B_i stand for a draw consisting of a white ball and $\neg\mathsf{B}_i$ for a black one at attempt *i*. Then for arbitrary knowledge K compatible with the 'rules of the game' we have

$$(*)\qquad \begin{aligned} &\mathfrak{w}(\mathsf{B}_i, \mathsf{K} \wedge \mathsf{A}_1) = 2/3, \quad \mathfrak{w}(\neg\mathsf{B}_i, \mathsf{K} \wedge \mathsf{A}_1) = 1/3 \\ &\mathfrak{w}(\mathsf{B}_i, \mathsf{K} \wedge \mathsf{A}_2) = 1/3 \quad \mathfrak{w}(\neg\mathsf{B}_i, \mathsf{K} \wedge \mathsf{A}_2) = 2/3 \end{aligned}$$

Let us analyse the judgment that normally leads us, after a fairly large number of observations, to the conclusion:

This is 'certainly' the first, resp. second, urn.

According to W9 we have, as in the proof of R6, with values out of (*)

$$(1)\qquad \frac{\mathfrak{w}(\mathsf{A}_1, \mathsf{K} \wedge \mathsf{B}_i)}{\mathfrak{w}(\mathsf{A}_2, \mathsf{K} \wedge \mathsf{B}_i)} = \frac{\mathfrak{w}(\mathsf{A}_1, \mathsf{K})}{\mathfrak{w}(\mathsf{A}_2, \mathsf{K})} \cdot \frac{\mathfrak{w}(\mathsf{B}_i, \mathsf{K} \wedge \mathsf{A}_1)}{\mathfrak{w}(\mathsf{B}_i, \mathsf{K} \wedge \mathsf{A}_2)}$$

$$(1')\qquad = \frac{\mathfrak{w}(\mathsf{A}_1, \mathsf{K})}{\mathfrak{w}(\mathsf{A}_2, \mathsf{K})} \cdot \frac{2}{1}$$

$$(2)\qquad \frac{\mathfrak{w}(\mathsf{A}_1, \mathsf{K} \wedge \neg\mathsf{B}_i)}{\mathfrak{w}(\mathsf{A}_2, \mathsf{K} \wedge \neg\mathsf{B}_i)} = \frac{\mathfrak{w}(\mathsf{A}_1, \mathsf{K})}{\mathfrak{w}(\mathsf{A}_2, \mathsf{K})} \cdot \frac{\mathfrak{w}(\neg\mathsf{B}_i, \mathsf{K} \wedge \mathsf{A}_1)}{\mathfrak{w}(\neg\mathsf{B}_i, \mathsf{K} \wedge \mathsf{A}_2)}$$

$$(2')\qquad = \frac{\mathfrak{w}(\mathsf{A}_1, \mathsf{K})}{\mathfrak{w}(\mathsf{A}_2, \mathsf{K})} \cdot \frac{1}{2}$$

Since in (1′) and (2′) K stands for arbitrary knowledge already supplemented by preceding observations, the following holds (with B^* describing a sequence of observations consisting of *w* white and *b* black draws (in arbitrary order)):

$$(3)\qquad \frac{\mathfrak{w}(\mathsf{A}_1, \mathsf{K}_0 \wedge \mathsf{B}^*)}{\mathfrak{w}(\mathsf{A}_2, \mathsf{K}_0 \wedge \mathsf{B}^*)} = \frac{\mathfrak{w}(\mathsf{A}_1, \mathsf{K}_0)}{\mathfrak{w}(\mathsf{A}_2, \mathsf{K}_0)} \cdot \left(\frac{2}{1}\right)^w \cdot \left(\frac{1}{2}\right)^b$$

$$= \frac{\mathfrak{w}(\mathsf{A}_1, \mathsf{K}_0)}{\mathfrak{w}(\mathsf{A}_2, \mathsf{K}_0)} \cdot 2^{w-b}\ [48]$$

But on account of $\mathsf{K}_0 \vDash \mathsf{A}_1 \leftrightarrow \neg\mathsf{A}_2$, we have

$$\begin{aligned} &\mathfrak{w}(\mathsf{A}_1, \mathsf{K}_0) + \mathfrak{w}(\mathsf{A}_2, \mathsf{K}_0) \\ &= \mathfrak{w}(\mathsf{A}_1, \mathsf{K}_0 \wedge \mathsf{B}^*) + \mathfrak{w}(\mathsf{A}_2, \mathsf{K}_0 \wedge \mathsf{B}^*) = 1, \end{aligned}$$

and (3) therefore shows how (e.g.) $\mathfrak{w}(\mathsf{A}_1, \mathsf{K}_0 \wedge \mathsf{B}^*)$ is determined by

$\mathfrak{w}(A_1, K_0)$. Thus apart from the *initial bias* which is expressed in the value of $\mathfrak{w}(A_1, K_0)$, and thus determines $\mathfrak{w}(A_1, K_0 \wedge B^*)$, there is also a *decision* as to that deviation from 1 (for $\mathfrak{w}(A_1, K_0 \wedge B^*)$) within which A_1 may be regarded as 'practically certain'.

For some observations of $w-b$ in the special case of an equal distribution,[49] i.e., where $\mathfrak{w}(A_1, K_0) = \mathfrak{w}(A_2, K_0) = 1/2$, the degrees of confirmation are given in the following table:

$w-b$	$\mathfrak{w}(A_1, \ldots)$	$\mathfrak{w}(A_2, \ldots)$
1	0.67	0.33
2	0.80	0.20
3	0.89	0.11
4	0.941	0.059
5	0.970	0.030
6	0.985	0.015
10	0.999 02	0.000 98
13	0.999 88	0.000 12
17	0.999 992 4	0.000 007 6
20	0.999 999 05	0.000 000 95
30	0.999 999 999 07	0.000 000 000 93

It should be noted that an initial bias deviating from an equal distribution could be expressed in terms of black and white balls. The first trial sequences might then decrease the amount of information.

A more extensive investigation would show that for every initial bias – expressed in a non-sclerodox initial distribution and a certainty requirement – it is as certain as required that a sufficiently long series of trials will as certainly as required indicate the present urn to be present, so that our experience with the required certainty eventually leads us to a judgment whose content, though admittedly not its degree of certainty, is independent of our initial bias. Those few confronted with 'wrong' sequences of experience will go crazy and can be thought of as the 'victims of statistics'.

However, if we attempt to transfer this reasoning from the two-urn model to the case of deciding between two (e.g. physical) theories, we find that the proposition $A_1 \leftrightarrow \neg A_2$ (which now asserts that precisely one of the theories in question is correct) is itself in the nature of a theory and can therefore be included only with reservations among the items of knowledge K_0. Apart from this it would seem that our example differs from practical cases only in that in general there is a greater number of theories under discussion (e.g., $A_1, \ldots, A_m$ instead of A_1, A_2) and there are more observable events to be considered (e.g., $B_1, \ldots, B_n$ instead of $B, \neg B$). If we know the schema of values $\mathfrak{w}(B_k, K \wedge A_i)$ corresponding to (*) (p. 163), then we calculate the distribution correction (as above by means of W9) out of the $\mathfrak{w}(A_i, \ldots)$ on the basis of the observation described by B_k, i.e.:

$$\frac{\mathfrak{w}(A_i, K \wedge B_k)}{\mathfrak{w}(A_j, K \wedge B_k)} = \frac{\mathfrak{w}(A_i, K)}{\mathfrak{w}(A_j, K)} \cdot \frac{\mathfrak{w}(B_k, K \wedge A_i)}{\mathfrak{w}(B_k, K \wedge A_j)}$$

where, of course, we now have $\mathfrak{w}(A_1, \ldots) + \ldots + \mathfrak{w}(A_m, \ldots) = 1$.[50]

Let us illustrate this by two examples:

(1) Let A_0–A_{100} stand for assumptions to the effect that in an urn containing 100 (white or black) balls, the number of white ones is precisely that stated by the index. Then a sufficiently long series of attempts with an unknown one of these urns will 'eventually' assign the highest degree of certainty to the correct one. If we now drop the simplifying restriction of a fixed number of balls',[51] we obtain correspondingly high degrees of certainty for the propensity p 'effective' in the series of trials being within a prescribed neighbourhood of the observed ratio $w/(w+b)$. Then a sufficiently long series of trials may be regarded as a measurement of p with the result $w/(w+b)$, in which case, incidentally, the certainty reached will eventually be (largely) independent of the initial bias.

(b) Let A_1 and A_2 be physical theories that yield different numerical values a resp. b (with $a > b$) for a measurable quantity; and let $a-b$ be smaller than possible errors of measurement, so that the theories cannot be precisely distinguished on the basis of measurements of this quantity. Further, let the precision of measurement be independent of whether A_1 or A_2 'holds', and let it be given by the standard error s (on the usual assumption about the distribution of possible errors of measurement). Then one measurement (described by M_x) with the result x, yields the correction of the degree of certainty implied by the following ratio correction:

$$\frac{\mathfrak{w}(A_1, K \wedge M_x)}{\mathfrak{w}(A_2, K \wedge M_x)} = \frac{\mathfrak{w}(A_1, K)}{\mathfrak{w}(A_2, K)} \cdot Q$$

with[52] $Q = e^{\frac{a-b}{s} \cdot \left(\frac{x}{s} - \frac{a+b}{2s}\right)} = \left(e^{\frac{a-b}{s^2}}\right)^{\left(x - \frac{a+b}{2}\right)}$.

The first form of Q shows that it is a question merely of *ratios* of the theoretical values (a, b) and the measured ones (x) *to the standard error s.* In its second form Q indicates that (because $a > b$) every measured $x > \frac{a+b}{2}$ is evidence for A_1, every $x < \frac{a+b}{2}$ for A_2 – and by how much.

Thus when we apply rules like (a) and (b), our commitment to an assertion or a theory is always dependent upon certain previous decisions (as to the degree of certainty required and the nature of the initial distribution). Perhaps we should take this as indicating that – except in borderline cases – 'logical thinking' can only prepare the ground for decisions but cannot replace them.

NOTES

1 Cf., for example, Rosser and Turquette [1].
2 See Lewis–Langford [1].
3 The idea is Popper's [1]. His term reflects a propensity to happen at different degrees of certainty. There are attempts to understand the measured certainty as a more basic concept, of which those based on propensities only are best understood instances. Carnap's [1] term 'degree of confirmation' reflects more a dependence on supposed or explicitly given evidence than the 'nature' of the values.
4 Exactly as in the case of the original standard of measure, this convention presupposes specific empirical knowledge, which we might express in idealized form in the following propositions:
(1) Experimental conditions of this kind yield degrees of certainty;
(2) These degrees of certainty depend only on the numerical ratios.
We must here dispense with the question whether the experimental conditions are 'ideally realizable', since it needs a developed mathematical error theory.
5 The two formulations are equally justified in principle, but on the whole $w: (w+b)$ (i.e., for the general case: the ratio of the number of cases of one kind to the total number) yields simpler formulas; cf., for example, p. 167, note 11.
6 It is no doubt more than merely fortuitous that the extension of the concept of assignment should here parallel the (intuitive) interpretation of formulas in terms of events (cf. p. 108, note 24).
7 For counter-arguments, cf. p. 143.
8 Conversely: once these laws have been validated for arbitrary degrees of certainty, they can, of course, also be applied to propensities; cf., for example, VIII 2, (g), p. 152 f.

[9] This is a relation – usually symbolised by '$\leqq$' – with the properties $x \leqq x, x \leqq y \wedge y \leqq z \rightarrow x \leqq z, x \leqq z \vee z \leqq x, x \leqq y \wedge y \leqq x \rightarrow x = y$.
[10] Without this subsidiary condition it would be easy to find counter-examples.
[11] The details of this analysis would show what properties must here be presupposed for the scale of the abstract degrees of certainty. The choice of $x+y$ is, however, not cogent. Our initial simple choice above of $w : b$ instead of $w : (w+b)$, would here be matched by the much clumsier

$$\frac{x+y+2xy}{1-x \cdot y},$$

resulting from the immanent transformation of the scales (0 to 1, respectively 0 to infinity).
[12] We have written '*A* ...', instead of 'A ...' in order to indicate that these propositions are here to be regarded as undecomposable, i.e. as variables. But, if there is something 'variable' here, it is the state of that supposed world.
[13] This depends on the simplifying assumption that only a finite number of states are possible. In the general case only suitable sets of states could be weighted, or given a measure.
[14] The question whether distributions of distributions would be more adequate here will have to be left undiscussed for the simple reason, among others, that if we introduced them, we should have to give up our restriction to the finite to an even greater extent than we have already done by introducing distributions.
[15] This is a properly inadmissible simplification; strictly, we should write '$x_i|^2$', so as to include the case of complex coefficients.
[16] One difference between C_2, D and C_1, D is that with $\mathfrak{W}_A(B) = 0$, we have on the one hand $\mathfrak{W}_{C_2}(D) = 0$ but on the other hand $\mathfrak{W}_{C_1}(D) \neq 0$.
[17] The question whether quantum physics needs or suggests a non-classical logic is still controversial. For instance, *pro* see Suppes [1], *contra* see Fine [1].
[18] For which e.g. $\mathfrak{W}(A \wedge B) = \mathfrak{W}(A) \cdot \mathfrak{W}_A(B)$ would not hold generally.
[19] Cf. in this connexion VIII 4, p. 163 f.
[20] $\log_2 x$ is the number y with $2^y = x$. This can be simply calculated with the aid of a logarithm table, using the formula $y = (\log x) : (\log 2)$.
[21] Such weighted means, known in probability theory as 'expectations', play an important part in the theory.
[22] Presumably the word 'probability' goes back to the conception which we have expressed in 'degree of certainty', but it has today become so overlaid with connotations based on frequency interpretations that two concepts have had to be distinguished; thus Carnap's *probability*$_1$ (degree of confirmation, likelihood) and *probability*$_2$ (relative frequency). Cf. Carnap [1], p. 25 ff. This distinction may be accepted as illuminating if *probability*$_2$ is understood in a sense of propensity more closely related to frequency phenomena than the more general idea of certainty.
[23] Cf. VIII 1, p. 140. If the reader is familiar with the concepts of modern algebra, he may wish to formulate this more precisely.
[24] For objections to this inference, see above p. 146.
[25] Cf. in this connexion VIII 4, p. 164.
[26] E.g.: 'There is such a thing as telepathy', 'There are flying saucers'. Correction in the light of experience should yield appropriate degrees of certainty even for contested propositions such as these, insofar as there are no limits set to our calculations by the complexity of the empirical evidence (which would include the credibility of witnesses).

[27] We cannot here discuss the question whether the choice of the same scale conceals a real difference or avoids making an unreal one. For a possible approach, cf. VIII 2 (g), p. 152 ff.
[28] This 'conditioned probability' $\mathfrak{W}_{B}(A)$ must not be confused with $\mathfrak{W}(B \rightarrow A)$. Cf. also p. 144, and note 18.
[29] This formulation presupposes that no denominator is 0; similarly in some other cases. Though here the technically adequate handling of the borderline cases is a matter of simple algebra, there is a problem: the extension to richer languages seems to necessitate the introduction of infinitely small values different from 0, hence a 'non-archimedean' scale of degrees.
[30] Carnap [1], p. 285: '... accepted in practically all modern theories of probability'. Jeffreys [1] has a similar reduction of W7 to W5, which is introduced there as an extrapolation of provable cases.
[31] Which express what is known as Bayes's Theorem; for an application cf. below pp. 158, 160, 163.
[32] The reader should keep in mind that urns and dice here merely serve as examples for a general case.
[33] This is a typical problem of analysis in probability theory, the solution of which goes back to Jacob Bernoulli [1].
[34] Combinatorics tells us that there are

$$\frac{n \cdot (n-1) \cdots (n-g+1)}{1 \cdot \ 2 \ \cdots \ \cdot g}$$

different ways of doing this.
[35] Our table, calculated with the use of approximating formulas (error function with $h = 60$, see for instance Jeffreys [1], p. 72), holds only for $p = 1:6$; it is, however, typical.
[36] These ratios, that correspond to the possibility discussed above (p. 138), presumably form the basis for formulations such as 'a high degree of certainty' for degrees close to 1.
[37] Perhaps our choice of a scale for degrees of certainty (0 to 1) needs the justification of the fact that it makes the link with relative frequencies particularly easy.
[38] Cf. von Mises [1], p. 186.
[39] There are many transitional stages between the formation of conditioned reflexes and learning from experience.
[40] Cf. VI 2, p. 115 f.
[41] Thus, e.g., usually in the case of learning from experience.
[42] That is, without reference to degrees of any kind, with which calculations could be performed.
[43] Cf. our remarks on idealization in I 1, p. 11 and III 3, p. 62 f. A more precise analysis would have to show whether R1* can be applied even in cases where the lack of confirmation of the probability statements depends not (only) on the incompleteness of the available information, but (also) on the indeterminacy of concepts.
[44] For if $K, A \models B$, then: $K \models B$ if and only if $K, \neg A \models B$.
[45] Cf. above, note 29.
[46] The reader is reminded that such an analysis is possible only on the highly ideal condition of the 'world' to be described being capable of only a finite number of states.
[47] To prepare the way for more general formulations let A_1 again stand for A, A_2 for $\neg A$; and let it be assumed that $A_1 \leftrightarrow \neg A_2$ is included among the initial knowledge K_0. Note that, in general, such an assumption could reduce the remaining possibilities

to an infinitely small amount, such that only degrees in that wider sense mentioned in note 29 could yield a well-defined quotient.

48 The mere relevance of the *difference* is, of course, a peculiarity of our example. It is chosen for this simplicity; but it could be understood as another kind of norm: for (still) *two* cases (urns) to be compared, but *general* values in the schema (*) viz p, $1-p$, resp. q, $1-q$, the exponent (also to be applied in the subsequent table for $w-b$) is

$$w \cdot \log_2(p:q) + b \cdot \log_2(1-p):(1-q).$$

49 On the subject of 'equal distribution' cf. VIII 1, p. 146; 2, p. 147 f.

50 In the case where for some i, k $\mathfrak{w}(\mathsf{B}_k, \mathsf{K} \wedge \mathsf{A}_i)$ is zero, the appurtenant A_i is, of course, excluded by an observation B_k. Such borderline cases are automatically covered by the conventional mathematical formulation.

51 This simplifying restriction amounts, after all, to a sclerodox initial judgment about the possible ratios.

52 We here have $e = 2.71828\ldots$, which results from the quotient of two terms for the normal error law. Since $a-b > 0$, we have $e^{(a-b)/s^2} > 1$. In order to make this correction comparable with formula (3) p. 163, we could further write

$$Q = 2^{\frac{a-b}{0.693 \cdot s^2}\left(x - \frac{a+b}{2}\right)}$$

where 0.693 ... is due to the change of bases (from e to 2.)

BIBLIOGRAPHY

ABBREVIATIONS

JSL = *The Journal of Symbolic Logic*, 1936ff. (This contains, apart from original contributions, a bibliography of all relevant publications, as well as book reviews.)

Studies = *Studies in Logic and the Foundations of Mathematics*, North-Holland Publishing Co., Amsterdam. (Since 1951 there have appeared in this series comprehensive expositions, as well as reports on research in progress.)

Apart from titles referred to in the text, we cite below also a number of books that could serve as introduction to the topics discussed.

Ackermann, W., [1] 'Zur Axiomatik der Mengenlehre', *Math. Annalen* **131** (1956) pp. 336–345.

Ackermann, W., [2] 'Solvable Cases of the Decision Problem', *Studies* 1954.

Aebi, M., [1] *Kants Begründung der 'Deutschen Philosophie'*, Basel, 1947.

Bachmann, H., [1] *Transfinite Zahlen*, Berlin-Göttingen-Heidelberg, 1955, ²1967.

Bendiek, H., OFM, [1] 'Zur logischen Struktur der Gottesbeweise (I)', *Franziskanische Studien* **38** (1956) pp. 1–38.

Bernays, P. and Fraenkel, A. A., [1] 'Axiomatic Set Theory', *Studies* 1958.

Bernoulli, Jacob, [1] *Ars conjectandi opus posthumum*. Selected translation in Newman [3], pp. 1452–1455.

Beth, E. W., [1] 'The Foundations of Mathematics', *Studies* ²1965.

Bocheński, J. M., [1] 'Ancient Formal Logic', *Studies* ²1957. (Written from the viewpoint of modern symbolic logic).

Cantor, G., [1] *Contributions to the Founding of the Theory of Transfinite Numbers*. Translated, and provided with an Introduction and Notes, by Philip E. B. Jourdain. Dover reprint, New York, 1952. (German original in *Mathematische Annalen*, Vol. **46** (1895) and Vol. **49** (1897)).

Carnap, R., [1] *Logical Foundations of Probability*, Chicago, 1950.

Carnap, R., [2] *The Continuum of Inductive Methods*, Chicago, 1952.

Church, A., [1] 'An Unsolvable Problem of Elementary Number Theory', *Amer. J. Math.* **58** (1936) pp. 345–363.

Church, A., [2] 'A Note on the Entscheidungsproblem + Correction', *JSL* **1** (1936) 40f., 101f.

Church, A., [3] 'A Formulation of the Simple Theory of Types', *JSL* **5** (1940) pp. 56–68.

Cogan, E. J., [1] 'A Formulation of the Theory of Sets from the Point of View of Combinatory Logic', *Zeitschrift f. math. Logik u. Grundl. d. Math.* **1** (1955), pp. 198–240.

Curry, H. B. and Feys, R., [1] 'Combinatory Logic I', *Studies* 1958.

Davis, M., [1] *Computability and Unsolvability*, New York-Toronto-London, 1958.

Davis, M., [2] *The Undecidable*. Basic papers on Undecidable Propositions, Unsolvable

Problems and Computable Functions (ed. by M. Davis), Raven, New York, 1965.
Döhmann, K., [1] 'Die sprachliche Darstellung der aussagenlogischen Funktoren', *Logique et analyse* **2** (1959), pp. 68–98.
Fine, A. I., [1] 'Logic, Probability and Quantum Theory', *Philosophy of Science* Vol. **35** (1968), pp. 101–111.
Fraenkel, A. A., [1] 'Foundations of Set Theory', *Studies* 1958. See also Bernays-Fraenkel.
Frege, G., [1] *Philosophical Writings*, translations from the —, by P. Geach and M. Black, Oxford, 1960. Containing i.a. *Begriffsschrift* (Chapter I), Function and Concept.
Frege, G., [2] *The Foundations of Arithmetic*. Bilingual translation by J. L. Austin. Oxford, ²1953, reprinted 1959.
Frege, G., [3] *The Basic Laws of Arithmetic*. Translated by Montgomery Furth. Berkeley and Los Angeles, 1964. See also van Heijenoort [1]
Gentzen, G., [1] 'On the Existence of Independent Axiom Systems for Infinite Sentence Systems'; in [3], pp. 29–52. German original in *Math. Annalen* Vol. **107** (1932).
Gentzen, G., [2] 'Investigations into Logical Deduction', in [3], pp. 68–131 German original in *Math. Zeitschrift* Vol. **39** (1935).
Gentzen, G., [3] *The Collected Papers of* —, (ed. by M. E. Szabo), *Studies* 1969.
Gödel, K., [1] 'The Completeness of the Axioms of the Functional Calculus of Logic'; in van Heijenoort [1], pp. 582–591. German original in *Monatshefte Math. Phys.* **37** (1930).
Gödel, K., [2] *On Formally Undecidable Propositions of Principia Mathematica and Related Systems*. (1) Translated by Meltzer, with Introduction by Braithwaite, Edinburgh-London, 1962. (2) Translated by E. Mendelson in Davis [2], pp. 5–38. (3) Translated by van Heijenoort in Heijenoort [1], pp. 596–616. German original in *Monatsh. Math. Phys.* Vol. **38** (1931).
Gödel, K., [3] *On Undecidable Propositions of Formal Mathematical Systems*. Mimeographed. Institute for Advanced Study, Princeton, N.J., 1934. Reprinted in Davis [2], pp. 39–74.
Gumin, H. and Hermes, H., [1] 'Die Soundness des Prädikatenkalküls auf der Basis der Quineschen Regeln', *Archiv für math. Logik und Grundlagenforschung* **2** (1956) pp. 68–77.
Halmos, P. R., [1] *Naive Set Theory*, Princeton N.J., 1960.
Hailperin, Th., [1] 'Quantification Theory and Empty Individual Domains', *JSL* **18** (1953) pp. 197–200.
Heisenberg, W., [1] *Der Teil und das Ganze*, Gespräche im Umkreis der Atomphysik, Munich, 1969.
Hermes, H., [1] *Enumerability, Decidability, Computability*. Berlin-Heidelberg-New York, ²1969. (Translation from the German 2nd edition 1965).
Heyting, A., [1] 'Intuitionism. An Introduction', *Studies* 1965.
Hilbert, D. and Ackermann, W., [1] *Principles of Mathematical Logic*. Chelsea Publ. Co., New York, 1950. (After the German ed. Berlin, ²1938).
Hilbert, D. und Bernays, P., [1] *Grundlagen der Mathematik*, Vol. I, Berlin, 1934, 2nd ed. Berlin-Heidelberg-New York, 1968. Vol. II, Berlin, 1939, 2nd ed. Berlin-Heidelberg-New York, 1970.
Jeffreys, H., [1] *Theory of Probability*, Oxford, 1939, ³1961 (An extended theory of probability$_1$.)
Karp, Carol, [1] 'Languages with Expressions of Infinite Length', *Studies* 1964.

Kleene, S. C., [1] 'General Recursive Functions of Natural Numbers', *Math. Annalen* **112** (1936), pp. 727–742.
Kleene, S. C., [2] 'On Notation for Ordinal Numbers', *JSL* **3** (1938), pp. 150–155.
Kleene, S. C., [3] *Introduction to Metamathematics*, Amsterdam-Groningen, ²1959.
Kneale, W. and Kneale, M., [1] *The Development of Logic*, Oxford, 1962.
Kolmogorov, A. N., [1] *Foundations of the Theory of Probability*, Chelsea Publ. Co., New York, 1954 (2nd English ed.).
Kyburg, H. E., [1] *Probability and the Logic of Rational Belief*, Middletown, Connecticut, 1961.
Lewis, C. I. and Langford, C. H., [1] *Symbolic Logic*, New York, 1932, reprint New York, 1959. (Particularly for modal logic).
Lorenzen, P., [1] *Einführung in die operative Logik und Mathematik*, Berlin-Göttingen-Heidelberg, ²1969.
Łukasiewicz, J., [1] *Aristotle's Syllogistic from the Standpoint of Modern Formal Logic*, Oxford, ²1957.
Markov A. A., [1] *Theory of Algorithms*, Jerusalem, 1962. (Orig. Russian: Moskva-Leningrad, 1954).
Mates, B., [1] *Stoic Logic*, Berkeley-Los Angeles, 1953.
von Mises, R., [1] *Wahrscheinlichkeitsrechnung*, Leipzig-Wien, 1931.
Mostowski, A., [1] 'On Definable Sets of Positive Integers', *Fundamenta Math.* **34** (1947), pp. 81–112.
Nagel, E. and Newman, J. R., [1] *Gödel's Proof*, New York University Press, ²1960. (An introduction to Gödel [2]. A variant is in Newman [3], pp. 1668–1695.)
Newman, J. R., [1], [2], [3], [4] *The World of Mathematics*. An anthology in four volumes. ([3] and [4], in particular, contain several relevant contributions.)
Patzig, G., [1] *Aristotle's Theory of the Syllogism*. A Logico-Philosophical Study of Book *A* of the *Prior Analytics*, D. Reidel, Dordrecht, 1968.
Pólya, G., [1] *How to Solve It*, Princeton, ⁵1948.
Pólya, G., [2] *Patterns of Plausible Inference* (Vol. 2 of: Mathematics and Plausible Reasoning), Princeton N.J., 1954.
Popper, K., [1] 'The Propensity Interpretation of Probability and the Quantum Theory', pp. 65–70 and 88f in: *Observation and Interpretation*, (ed. by S. Körner), 1957.
Post, E. L., [1] 'Finite Combinatory Processes-Formulation I', *JSL* **1** (1936) pp. 103–105.
Post, E. L., [2] 'Formal Reductions of the General Combinatorial Decision Problem', *American J. Math.* **65** (1943) pp. 197–215.
Quine, W. V., [1] *From a Logical Point of View – 9 Logico-Philosophical Essays*, Cambridge, Massachusetts, 1953.
Quine, W. V., [2] *Mathematical Logic*, Revised edition, Cambridge, Mass., ²1955.
Reichenbach, H., [1] 'The Syllogism Revised', *Philosophy of Science* **19** (1952), pp. 1–16.
Rescher, N., [1] *Topics in Philosophical Logic*, D. Reidel, Dordrecht, 1968.
Rescher, N., [2] *Many-Valued Logic*, McGraw Hill Co., 1969.
(Both books also for modal and probabilistic logics.)
Russell, B., [1] *Introduction to Mathematical Philosophy*, London, ⁶1948 (chap. XVI for descriptions). See also Whitehead-Russell.
Savage, L. J., [1] *The Foundations of Statistical Inference*. London-New York, 1962. (A discussion on the applications of subjective probability to statistics.)

Schneider, H., [1] 'Semantics of the Predicate Calculus with Identity and the Validity in the Empty Individual Domain', *Portugaliae Mathematica* **17** (1958) pp. 85–96.

Scholz, H. und Hasenjaeger, G., [1] *Grundzüge der mathematischen Logik*, Berlin-Göttingen-Heidelberg, 1961.

Schütte, K., [1] *Vollständige Systeme modaler und intuitionistischer Logik*. Berlin-Heidelberg-New York, 1968.

Shannon, C. E. and Weaver, W., [1] *The Mathematical Theory of Communication*, Urbana, Ill., 1963. (For the concept and theory of information.)

Skolem, Th., [1] 'Peano's Axioms and Models of Arithmetic', in: Mathematical Interpretation of Formal Systems. *Studies* 1955. Also in [2]. (On the non-characterizability of the number series by first order axiom systems).

Skolem, Th., [2] *Selected Works in Logic*, (ed. by E. Fenstad), Universitetsferlaget, Oslo, 1970 (Contains some essential contributions to predicate logic, set theory and recursive functions).

Smullyan, R. M., [1] *Theory of Formal Systems*. Revised Edition, Princeton N.J., 1961.

Suppes, P., [1] *Introduction to Logic*. Princeton N.J., 1957, ^{5}1962. (A textbook for a first course in modern logic.)

Suppes, P., [2] 'The Probabilistic Argument for a Non-Classical Logic of Quantum Mechanics', *Philosophy of Science* **33** (1966) pp. 14–21.

Surányi, J., [1] *Reduktionstheorie des Entscheidungsproblems im Prädikatenkalkül der ersten Stufe*, Budapest, 1959.

Tarski, A., [1] *Foundations of the Calculus of Systems*, in [3], pp. 342–383. German original in *Fundamenta Math*. vol. **25** (1935) and vol. **26** (1936).

Tarski, A., [2] *The Concept of Truth in Formalized Languages*, in [3], pp. 152–278. Polish original in *Ruch Filisoficzny* vol. **12** (1930–1).

Tarski, A., [3] *Logic, Semantics, Metamathematics*, Oxford, 1956.

Tarski, A., Mostowski, A. and Robinson, R. M., [1] 'Undecidable Theories', *Studies* 1953.

Turing, A. M., [1] 'On Computable Numbers, with an Application to the Entscheidungsproblem'. *Proc. London Math. Soc. Ser.* 2, **42** (1936/37) pp. 230–265. 'A Correction' **43** (1937) pp. 544-546.

Van Heijenoort, J., [1] *From Frege to Gödel. A Source Book in Mathematical Logic, 1879–1931*, Cambridge, Mass., 1967.

Weingartner, P., [1] 'Modal Logics with Two Kinds of Necessity and Possibility', *Notre Dame Journal of Formal Logic*, vol. **9** (1968) pp. 97–159.

Whitehead, A. N. and Russell, B., [1]–[3] *Principia Mathematica*, 3 Vols. Cambridge, 1910–1913, ^{2}1925–27, reprint 1950.

INDEX

Besides places of definitions indicated by bold face page-numbers, occasionally other relevant pages are mentioned. Numbers in parentheses refer to passages relevant without showing the entry, or a variant. Upper indices at page-numbers identify notes. If need be, specification of field is indicated, particularly by: C (comprehensor), D (decision), F (function), I (identity), P (proposition), Q, Q^2 (quantification-, predicate), S (many-sorted), T (type), 𝔚 (probability), Σ (syllogistic), ▶ (sequence).

A. SYMBOLS, ALSO RELATED TO PART B

B. ALPHABETICAL PART

INDEX